Notice of Peer Review
Professor Charles Romesburg was an early adopter of electronic printing on demand, and this is his fourth book published in that way (Romesburg 2002, 2004, 2005). Printing on demand has some advantages over conventional publishing. There are no warehousing costs, and the text can be updated whenever needed. On the other hand, a commercial or academic publisher would provide additional services, including peer review. To ensure the academic quality of Professor Romesburg's book, the Department of Environment and Society, Utah State University, commissioned two independent reviews of the final book draft by qualified scholars. The reviews were coordinated by Professor James Kennedy of this department. Professor Kennedy selected reviewers who have no relationship with Professor Romesburg, or connection to Utah State University. The reviewers were asked to read the text for technical accuracy. No technical errors were found.

Professor Romesburg has published previously in the area of gaining scientific knowledge (Romesburg 1981, 1991), the topic he covers in this book. He has taught the subject to graduate and undergraduate students, as well as to technical specialists of the U.S. Forest Service. Professor Romesburg will use this book in his teaching, and we hope others find it useful as well.

Joseph A. Tainter
Department Head
Department of Environment and Society
Utah State University
Logan, Utah

Romesburg, H. Charles

1981   Wildlife Science: Gaining Reliable Knowledge. *Journal of Wildlife Management*, 45: 293-313.

1991   On Improving the Natural Resources and Environment Sciences. *Journal of Wildlife Management*, 55: 744-756.

2002   *The Life of the Creative Spirit.* Philadelphia: Xlibris Corp.

2004   *Cluster Analysis for Researchers.* Morrisville, NC: Lulu Press, Inc.

2005   *How About it, Writer.* Morrisville, NC: Lulu Press, Inc.

BEST RESEARCH PRACTICES

How to gain reliable knowledge

H. Charles Romesburg

ISBN 978-0-557-01783-6

Published by
Lulu Enterprises, Inc.
Morrisville, North Carolina
Website: www.lulu.com

Citation format: Romesburg, H. C. 2009. *Best Research Practices*. Morrisville, NC: Lulu Enterprises, Inc.

All of the author's royalties from this book are donated to no-kill animal shelters.

Author's weblog: www.comesaunter.com

The author dedicates this book to his mother, for bringing him up hooked on education.

Contents

Preface

The story of this book does not take long to tell. It begins early in my career while I was studying research practices in wildlife science. I realized that improving the practices promised to increase the reliability of wildlife knowledge. I explained this in an article, "Wildlife Science: Gaining Reliable Knowledge" (Romesburg 1981). The article struck a nerve among some wildlife researchers, and won a Wildlife Society publication award. I wrote follow-on articles (Romesburg 1993, 1991, 1989), as did others. (A good way to find them and learn of the discussions and debates they inspired is to Google on the Internet "reliable knowledge" paired with "Romesburg.")

From there grew my interest to develop material for this book. Over the last thirty years, I studied more than 5,000 top scientific articles. By top scientific articles, I mean they were published in the journal *Science* or *Nature*, or editors of other journals in which they appeared gave them special mention, or *Science News* brought them to notice. I studied the research practices that supported the articles' findings: the scientific methods used, and how they were used. Because they came from articles in top journals, I called them "best research practices." No collection can be complete, yet I did cast a wide net in compiling this one.

At the same time, I studied letters to the editors of top journals that pointed out misuses of scientific methods. And in articles in *Skeptical Inquirer* I studied the methods of those outside of science who produced bogus claims of knowledge but proclaimed them valid. By this I was studying the worst research practices. I learned that the worst research practices simply neglect to adhere to the best research practices.

The best research practices in this book apply to all fields: all of the physical sciences, life sciences (so including wildlife science), earth sciences, environmental sciences, psychological sciences, and social sciences. The book is designed for self-study. To get the most from it, you should understand the subject of inferential statistics. Yet even lacking that, practically the entire book is understandable.

What's in the book for you depends on which one of the following ten groups you belong to:

1. If you are a graduate student in science: Learning the best research practices when you enter graduate school will help you hit the ground running in your studies. You'll be reading research articles and listening to talks reporting research. The book stresses analyzing how the research was planned and carried out, and how possibly it could have been done better to increase the reliability of the findings. And the book should give you ideas and approaches for your research.

2. If you are a professor teaching graduate students: (a) If you teach a graduate readings course where the class analyzes and discusses leading research articles, consider starting the course off by having the students read this book. It will put them "on the same page." It will show

them how to pigeonhole research articles according to the goals addressed and methods used. (b) If you are the major professor of graduate students, you know that entering graduate students grope for two or three years before settling in on a thesis topic. Having them read this book early on should speed up their search, and result in better thesis topics ("better" meaning a better chance the research will eventually be well cited).

3. If you are a professor teaching undergraduates: (a) As assigned background reading, the book can be shoe-horned into any undergraduate science course, without causing you to get out of your groove. Science courses cover what is known. The book will give students an awareness of the methods by which what is known comes to be known. You can refer to parts of it when appropriate, as with a biology professor saying: "In arriving at his theory of natural selection, Darwin relied on the method of retroduction. Recall that retroduction is covered in chapter 3 of Professor Romesburg's book. Darwin couldn't use the hypothetico-deductive method, covered in chapter 4, to test his hypothesis because. . . ." (b) If you are a mentor to undergraduate researchers, having them study the book will help them design their research projects.

4. If you are a practicing scientist: The book's material has fallen through the cracks of university education. When you were in college you probably had only a smattering of it. The book should improve your skills for evaluating the reliability of knowledge claimed in research articles, letting you independently make up your mind. And the book's diverse set of research examples should aid border-crossing thinking, suggesting applications of scientific methods you may be unfamiliar with. For instance, an ecologist seeing an approach taken in medical research but not usually in ecology may get an idea for applying it in ecological research.

5. If you are a practicing manager, engineer, planner, or technologist: As a decision maker you need to assess the reliability of the published research you base your decisions on, rather than uncritically accepting the authors' judgments. For instance, environmental managers must evaluate the reliability of knowledge produced by environmental scientists. Knowing the simple fundamentals of best research practices helps make a correct evaluation.

(I have a suggestion for managers, engineers, planners, technologists, and scientists working in research departments of companies and governments: Consider discussing and debating ideas in the book during lunchtime brown-bag seminars.)

6. If you are a lawyer or judge who handles cases where scientific evidence plays a role: A good grasp of best research practices will help you distinguish "junk science" from trustworthy science.

7. If you are a student in an applied statistics program: In the program's practicum, you advise faculty and graduate students about the best statistical practices for their research projects. To do this well, you need to be up on the scientific methods that underpin their research. And the same advice holds if you are a practicing statistician, some years out of college.

8. If you teach or practice science journalism: Science journalism is devoted to explaining science discoveries to the general population, helping us all make sense of our world. Science journalists must have an encyclopedic grasp on the ways by which knowledge is produced, so that from interviews with scientists, and from reading journal articles, they can accurately portray how credible the claimed discoveries are. Thus, the book might belong on the syllabi of those teaching science journalism, and on the shelves of those practicing the profession.

9. If you teach critical thinking skills or want to improve yours: Researchers are among the best critical thinkers. By covering how researchers do basic research, this book incidentally covers critical thinking skills.

10. Consumers of any kind of information: The book should benefit non-scientists who want to know how to weigh the validity of the information they consume, and to analyze statements that are presented to them as verified by science but which might be empty talk.

From my teaching the methods of science for years, and seeing my students' course evaluations, I believe in the power of first learning the methods from everyday examples, such as applying the method of retroduction to solve a crime. When my students are comfortable with that, I give them examples of researchers applying the methods to learn about nature. This is why I have loaded the book with everyday examples and research examples. The research examples do four jobs:

(1) The research examples will remind you of what important knowledge is, as opposed to unimportant knowledge. It is not enough that knowledge be reliable. It must also hold promise of being important, consequential, built on and cited by future researchers, or of leading to living better. Or to cast this in the negative, discoveries must not be of small worth, research money poured down the drain. In selecting the research examples, I bet on my judgment that the future will count the research as valuable.

(2) The research examples will dispel any idea that some of the scientific methods are limited to certain fields. I don't want anyone left thinking, "Yes, I see how the hypothetico-deductive method [or some other method] works in physics, but I'm not sure it'll work in my field." To dash such notions, I chose the research examples from a wide variety of fields.

(3) The research examples will strengthen respect for the so-called "soft" sciences, like psychology, sociology, geography, anthropology, economics, and ecology – this in comparison with the "hard" sciences like physics and chemistry. There is a belief among some in the hard sciences that, by and large, research in the soft sciences must produce shaky conclusions. There are research examples in this book that will burst that belief.

(4) Having plenty of research examples allows me to inculcate certain ideas. For instance, the book's set of research examples illustrating the method of crucial experiments is (I believe) several times larger than any set ever published. Had I included just one or two, that might leave the idea that this is a minor scientific method; it's not. It is a major scientific method. Besides, the large set will go far toward impressing you with the possibility of using crucial experiments in

your research.

I have a story that summarizes the power of learning with examples. In my early years in my course on scientific methods, I taught the hypothetico-deductive (H-D) method with propositional logic, centered around the truth table for the equivalence relation (Kemeny 1959). The students' eyes glazed over. They never developed a natural intuitive understanding of the method. In time, I dropped the approach and began inventing everyday examples of the method, such as "the car example" and "the dam-river example" that you'll see in chapter 4. The everyday examples led to a key concept of the H-D method, which I named "shunt causes." By introducing the H-D method with everyday examples and then following with research examples, and drawing on the concept of shunt causes, my students got the method into their bones. You should too, in chapter 4.

As to the methods in this book, they are the prevalent ones used in basic research that is published in *Science* and *Nature*, as well as in other high-impact basic research journals (those most cited). Let me instance a staple of these journals: the frequentist methods of statistics. Naturally then, this book presents frequentist methods (as in chapter 2 and the Appendix). At the same time, methods that never (or rarely) appear in *Science* and *Nature* and other top basic research journals are absent from this book. Let me instance one such absence: the Bayesian methods of statistics.

If this were a movie, the credits would now roll. Yet rather than thanking hundreds of people individually, and risk forgetting some, I'll do it en masse. I am grateful to the students who have taken my best research practices course (and its predecessor course). Many of them have asked questions I hadn't considered, and answered questions I had considered but hadn't answered. And I thank the researchers who have written articles about gaining reliable knowledge. Their articles gave me some good ideas and encouraged me to continue working on the subject. And I want to thank in advance any science teachers who, realizing that the lack of a required course devoted to scientific methods is a longtime weakness in education, develop a course devoted to teaching best research practices, and reason with their faculty to make it required of graduate students and upper-level undergraduates.

1
Owner's manual for this book

As I like to think of it, it is easily possible that in a few years the average reliability of knowledge in research journals will be above its baseline level of today. This can happen through researchers making better use of scientific methods.

There is not one scientific method – not *the* scientific method. There are many. "Method" comes from the Greek *methodos*, meaning "pursuit of knowledge." *Methodos* is based on *hodos*, meaning "way." A scientific method is a way of pursuing knowledge. There are the main scientific methods, e.g., the hypothetico-deductive method, and the retroductive method. And there are varieties of the main methods, though rarely have scientists given them names. Some of the methods and their varieties are especially powerful at discovering truths. And some that aren't can be yoked together in especially powerful combinations.

This book is a compilation of the methods and their varieties. The first thing you should do is read it through and let the methods sink in. Then draw on them when you do research, and when you judge the reliability of published research. As I tell my students: "Give yourself some years of practice with the methods, and in time you and science will be rewarded."

For doing research: Among the goals that your research project might involve are the following: planning and running an experiment; conceiving and circumstantially supporting a research hypothesis; testing a research hypothesis; discovering the possible existence of cause and effect; and investigating the possibility that chance could easily have caused an interesting existing fact. Consider the goals in turn:

For planning and running experiments, follow chapter 2's advice for producing reliable data. When all is done, ask yourself: "Am I confident that the experimental measurements – the data – contain no more than a tolerably small amount of random error and a tolerably small amount of systematic error?" If "Yes," consider the data reliable. If "No," there's no excuse for it; reliable data are always attainable. Start over from square one and do the experiment right, or give up the project.

I realize I am saying this in stringent terms. I mean for you to not take my advice lightly. For, it is easy to produce unreliable data but not realize that you have. I call your attention to Ioannidis' (2005) article, "Why Most Published Research Findings Are False." Ioannidis places the cause on researchers who fail to follow best research practices in planning and carrying out their experiments. The measurements – the data, the facts – from their experiments are made intolerably uncertain by large random error and/or systematic error. Unaware this has happened, they believe the data are reliable; in reality, the data are wrong.

Before we leave Ioannidis, I have this to say: he analyzed medical research findings, and so there is no justification for extrapolating his claim of "most" to all fields. Doubtless the percentage of false findings in physics, for instance, is much less than in medicine. Second, Ioannidis' claim of "most" may be an exaggeration, as Goodman and Greenland (2007) believe. All the same, it is safe to say that of research in fields other than physics, chemistry, and the like, a substantial minority of published findings is false because of unreliable data.

More of chapter 2's space is spent on systematic error than on random error. It's not that systematic error is more important. Error is error, regardless of its kind. It's just that you already know much about random error from your statistics course(s). Education nearly ignores systematic error. I'm trying to right the imbalance a bit.

As you read beyond chapter 2, keep in mind that even when the data are reliable, the conclusions that researchers base on the data run a risk of being unreliable. An example: Suppose the year-by-year numbers of cricket frogs in Illinois, which show a decline, are known absolutely. Nonetheless, theories proposed to explain this absolute truth could be false. And keep in mind that research that is suggestive rather than conclusive is valuable. For, one or several related research findings, each with reservations, can sometimes be assembled into a critical group where each solves the others' reservations, resulting in reliable knowledge. This is why science journals publish research findings that are not true beyond a reasonable doubt. Such articles wait for other articles to be published whose findings likewise are not true beyond a reasonable doubt. In time, from the accumulated information a reliable conclusion often results. Let's go over this again in terms of the phrase "gaining reliable knowledge," which has a parallel in American football: Occasionally a research project comes along that gains the full length of the football field. Most research projects, however, grind out shorter gains of the four yard or twelve yard variety. The shorter gains wait for the possible opportunity of joining up with other short gains, producing a touchdown, a reliable conclusion.

The aim of the chapters beyond chapter 2 is to help researchers add a few more yards to their plays. Instead of a four-yard gain, get a nine-yard gain. Instead of a twelve-yard gain, get a seventeen-yard gain. With longer plays there is a greater promise for the yards gained to combine into a touchdown. And of course, the other aim is to increase the probability of advancing all the way in a single project, a hundred yard gain, reliable knowledge in one go.

For conceiving and supporting research hypotheses, chapter 3 will remind you of what to do. It will remind you to concurrently practice the three methods for conceiving research hypotheses: the retroductive method, the question method, and the subject method. It will remind you of the steps of developing support for the hypotheses. It will remind you to resist the temptation to prematurely stop the recursive process of conceiving and supporting hypotheses. It will remind you that smoking-gun circumstantial evidence and information can sometimes be discovered that will prove the hypothesis is true beyond all dispute.

For testing research hypotheses, consult chapter 4's coverage of the hypothetico-deductive (H-D) method and its varieties. Often, conceiving and supporting a research hypothesis is only able to move toward a reliable conclusion, and ends up short. Testing the hypothesis is often able to reach the rest of the way. And chapter 4 covers some ideas that aren't widely known. To mention one: Ways can usually be found to do H-D tests with relatively cheap experiments rather than expensive ones. To mention another: A thesis committee would shudder at the candidate who announced that she or he had tested a hypothesis using existing data, instead of with data collected after the hypothesis had been conceived. Yet as explained in chapter 4, under certain conditions this is perfectly ok.

For investigating the possible presence of cause and effect, go to chapter 5. It covers the two methods for discovering cause and effect: the direct method and the indirect method. Further, it explains why neither method is categorically superior for making reliable discoveries of cause and effect. In certain situations, the direct methods does better; in others, the indirect method.

For deciding whether chance could have easily caused an interesting observed fact, go to chapter 6. For whenever chance is probably the cause, it's a bad idea to spend time trying to come up with a scientific theory to explain the fact. Another thing, the methods for deciding whether chance is or isn't behind an existing fact have applications in everyday life. They can help the lay public distinguish between pseudoscience and legitimate science. Do dowsers have skill at finding underground water? Are there hidden codes in the Bible?

For how to build and use reasoning models, go to chapter 7. Reasoning models are a few steps of deductive logic, accomplished in minutes. Researchers make reasoning models for any of three purposes: (1) to predict facts, (2) to explain processes of nature, or (3) to do thought experiments.

For how to build and use computer simulation models, go to chapter 8. Computer simulation models are hundreds of steps of deductive logic, cast as a computer program, accomplished in weeks and months. Researchers make computer simulation models for the same three purposes that they make reasoning models: (1) to predict facts, (2) to explain processes of nature, or (3) to do thought experiments.

For how to minimize the ethical biases in research, go to chapter 9. Most of us have unconsciously-rooted biases about research, and unless we deliberately keep them at bay, progress in our fields will suffer. Our first step to prevent the biases from acting is to know what they are, and chapter 9 lists them.

For why the funding of basic research must be increased, go to chapter 10. Basic research is in trouble. It is not receiving from funding agencies the attention to which its crucial importance entitles it. It needs vocal supporters. What should they say? What chapter 10 says.

For tips on how to become a better researcher, go to chapter 11. Patiently practice them and, I feel sure, in time they will work magic on you, as they have for many.

For judging the reliability of published research: Let's start with a reminder of the obvious. The titles and abstracts of science articles state the conclusions – the findings, the claims of knowledge – drawn from research. But from just reading the conclusions there is no way of telling whether they are reliable or not. To demonstrate, I just pulled a half dozen articles from a pile in my office, and from their titles and abstracts extracted their conclusions.

1. Conclusion: Hypercold winters obscured a global trend of rising summer temperatures at the end of the last Ice Age (Schaefer et al. 2006). 2. Conclusion: Exposure to human pheromones affects the length of women's menstrual cycles (Stern and McClintock 1998). 3. Conclusion: Throughout geologic time, the tropics have been a major source of evolutionary novelty (Jablonski 1993). 4. Conclusion: There is a direct connection between smoking cigarettes and depression (Hemenway et al. 1993). 5. Conclusion: To minimize their energy consumption during migration, female butterflies will decrease their airspeed in a tailwind and increase it in a headwind (Srygley 2001). 6. Conclusion: Strong present-day aerosol cooling implies a hot future (Andreae et al. 2005).

I am not suggesting these conclusions are unreliable. I am demonstrating that it is impossible to tell just by looking at them. Thus, when you read a research article, you must analyze the body of the article, and from this estimate the reliability of the conclusion independently of the author's estimate. The information in chapter 2 will help you estimate the reliability of the experimental or observational data reported in the article. Given that you decide the data are reliable, the information beyond chapter 2 will help you estimate the reliability of the conclusions.

I leave you with a touchstone. Just as the profession of medicine has its Hippocratic Oath, the profession of science has its moral imperative. Jacob Bronowski (1965) states it: *"We OUGHT to act in such a way that what IS true can be verified to be so."*

2

How to plan and carry out reliable experiments

EXPERIMENTS! Experiments produce facts. Facts are the basis for making decisions. Facts are often the starting point for creating research hypotheses. Facts are always the ending point for testing research hypotheses. And for all their uses, facts must be reliable: acceptably free of systematic error and random error.

At an early age we learn about systematic error and random error, and the goal of minimizing them. I learned from my mother's taking my temperature with a mercury thermometer. To keep the effect of systematic error small, she would shake the mercury down into the bulb before starting the experiment, and have me keep the bulb under my tongue for three minutes. To keep the effect of random error small, she would repeat the experiment several times and mentally average the measurements.

Let's be clear about the terms:

Facts. The facts we are talking about are quantitative facts. As such, "facts," "measurements," and "data" mean the same.

Random error and systematic error. These terms are used in two senses: (1) Facts from an experiment contain random error and systematic error that shift the facts from their true values: in the case of random error, it bumps the facts off of their true values by a random amount; in the case of systematic error, it slides them away from true value by a fixed amount. (2) Experiments contain sources of random error and systematic error. By designing and executing a **reliable experiment** – one acceptably free of sources of random error and systematic error – **reliable facts** are produced – facts acceptably free of random error and systematic error.

The true value of a fact is the value it has in the absence of random error and systematic error. When we look at a fact, we do not see the true value of the fact. We see what has resulted from the true value being altered by random error and systematic error.

In this chapter I'd like to show you how to minimize the random error and systematic error in your experiments – minimize the sources of the two and minimize the sizes of the two – so your experiments will produce reliable facts. For this, I have organized the chapter into three parts.

1–SETTING THE STAGE

Section 2.1 illustrates random error and systematic error.

Section 2.2 explains the importance of minimizing systematic error in planning and carrying out experiments, making it as small as proves practically possible.

Section 2.3 explains the importance of protocols for minimizing systematic error.

2–COVERING SPECIFICS (THE HEART OF THIS CHAPTER)

Section 2.4 explains the stages in planning and carrying out experiments: the sampling stage, the pre-measurement stage, and the measurement and calculation stage.

Section 2.5 explains how to state in your research reports the sizes of the random error and the systematic error in your experiments.

Section 2.6 lists some common sources of systematic errors. Being familiar with them will help you avoid those that can be avoided in planning and carrying out experiments, and help you minimize those that can't be.

Section 2.7 explains how the pre-measurement stage of experiments is often where the most potentially damaging sources of systematic error lurk. That is where you should give extra attention at finding and minimizing them.

Section 2.8 explains why you should try to use standard constructs in the measurement stage of your experiments. One reason is that standard constructs minimize systematic error in making measurements.

Section 2.9 explains why systematic errors in comparative experiments, commonly thought to cancel out, don't always.

Section 2.10 explains a five-step method to minimize systematic error in experiments.

Section 2.11 explains how, with the aid of a collaborating statistician, to determine the proper sample size for minimizing random error in experiments.

3–INCLUDING IMPORTANT SIDE TOPICS

Section 2.12 offers advice on rewarding your collaborating statistician.

Section 2.13 covers experiments that do not involve sample units.

Section 2.14 covers three key questions that researchers describing their experiments must answer.

Section 2.15 explains induction – reasoning from knowledge about a few members of a class to knowledge about all members of the class – of which the methods of statistical inference are an instance.

Section 2.16 covers two kinds of qualitative facts.

Section 2.17 introduces P-plane diagrams and the place of experimental facts in them. P-plane diagrams are a way of picturing applications of the various methods of science.

We will cover the above material in terms of experiments for estimating quantitative facts. Such experiments are the bread and butter of research. They run from an experiment to estimate the mean uranium content in tektites scattered about the Libyan Desert; an experiment to estimate the mean number of micrograms of mercury per gram of albacore tuna; an experiment to estimate the mean number of bullfrogs in a region of Oregon; an experiment to estimate the mean level of airborne PM-2.5 on a given day in Cleveland – to an experiment to estimate the mean I.Q. of sixth graders in Toronto. At the same time, the material applies to other kinds of quantitative experiments we will not illustrate, such as analysis of variance experiments, randomized block experiments, split plot experiments, and factorial experiments.

As you read the chapter, you may notice that it gives less space to covering random error and how to minimize it than it gives to systematic error and how to minimize it. Don't let that mislead you into thinking that random error is a lesser threat to the goal of getting reliable data. It's just that random error and how to minimize it are well covered in courses and books on inferential statistics, while systematic error is very little discussed.

2.1 An illustration of random error and systematic error

Let's picture how random error randomly bumps measurements from their true value, and how systematic error slides measurements by a fixed amount from their true value. The picture consists of imagining a set of four experiments – A, B, C, and D – to estimate the weight of a certain apple. Nature knows the apple's true weight, $\mu = 135.17$ gm. We don't. For each experiment we weigh the apple 10,000 times on a scale, which is to say measure its weight 10,000 times, which is to say replicate (run) the experiment 10,000 times, which is to say that the sample size of each experiment is $n = 10,000$. Several of the experiments are free of random error or systematic error, or both, making them unrealistic but instructive.

Experiment A is errorless. It has no random error and no systematic error. We weight the apple the first time; its measured weight is $x_1 = 135.17$ gm, its true weight. We weight it the second time; its measured weight is $x_2 = 135.17$ gm, its true weight. It is the same for every time we weigh it, up through $x_{10,000}$.

For experiment A, I invite you to take a piece of paper and sketch the form of the frequency distribution, $f(x)$, of the 10,000 measurements. All of the measurements are piled up in a spike at the apple's true weight. The mean of the 10,000 is $\overline{X} = 135.17$ gm.

Experiment B has random error but no systematic error. Each time we run the experiment, the reading on the scale is randomly bumped off of the apple's true weight (perhaps from randomly fluctuating air currents in the room). Sometimes it is bumped higher, sometimes lower, and more often is bumped a little bit than a lot. In the 10,000 times we run the experiment, the measurements are, say, $x_1 = 134.15$ gm, $x_2 = 135.47$ gm, $x_3 = 136.11$ gm, $x_4 = 135.06$ gm, $x_5 = 133.98$ gm, $x_6 = 135.03$ gm, . . . , $x_{10,000} = 135.63$ gm. We add the 10,000 measurements and divide by 10,000, giving the mean $\overline{X} = 135.09$ gm. That's our estimate of the apple's true weight, $\mu = 135.17$ gm.

For experiment B, a sketch of the frequency distribution of the 10,000 replications is bell-shaped, the statistician's normal distribution, centered on the mean weight of $\overline{X} = 135.09$ gm, close to the apple's true weight of 135.17 gm. Why is the distribution bell-shaped? In making measurements, random errors are as likely to be positive as negative, and small random errors are more likely than larger ones.

Experiment C, like experiment B, has random error but no systematic error. But relative to B, the random error is greater. Consequently, the measurements tend to be bumped further from the apple's true weight. In the 10,000 times we run the experiment, the facts we measure are, say, $x_1 = 131.77$ gm, $x_2 = 138.64$ gm, $x_3 = 135.32$ gm, $x_4 = 137.00$ gm, $x_5 = 136.31$ gm, $x_6 = 134.03$ gm, . . . , $x_{10,000} = 135.45$ gm. You can see they vary more than those in experiment B. We add the 10,000 and divide by 10,000; the mean is $\overline{X} = 135.19$ gm.

For experiment C, a sketch of the frequency distribution of the 10,000 runs is bell-shaped, centered on the mean weight of $\overline{X} = 135.19$ gm, close to the apple's true weight, 135.17 gm. The bell-shaped distribution is spread out more than that for experiment B because the random error is more variable. In statistical terms, the standard deviation for the distribution of measurements produced by experiment C is larger than the standard deviation for the distribution produced by experiment B.

Experiment D is like experiment B in the random error it has, but unlike B the weighing scale is miscalibrated, producing a systematic error in the measurements, a negative one of size 4.72 gm. Each time we run this experiment, the random error bumps the measurement off of true value, randomly higher or lower, as with experiment B. At the same time, the bumped measurement is slid lower by 4.72 gm because of the systematic error. In the 10,000 times we run the experiment, the measurements are, say, $x_1 = 130.77$ gm, $x_2 = 129.41$ gm, $x_3 = 130.91$ gm, $x_4 = 131.41$ gm, $x_5 = 130.28$ gm, $x_6 = 129.31$ gm, . . . , $x_{10,000} = 129.98$ gm. We add the 10,000 measurements and divide by 10,000; the mean is $\overline{X} = 130.36$ gm.

For experiment D, a sketch of the frequency distribution of the 10,000 replications is bell-shaped, with essentially the same spread on the bell shape as that of experiment B. But it is centered about 4.72 gm lower than the apple's true weight. Without the systematic error, the mean of the 10,000 replications would be $\overline{X} = 135.08$ gm (130.36 gm + 4.72 gm). The systematic error has slid this 4.72 gm lower, giving $\overline{X} = 130.36$ gm (135.08 gm - 4.72 gm).

The above experiments consist of a measurement and calculation stage. But what they illustrate about random and systematic errors applies to real research experiments. Besides the measurement and calculation stage, real experiments often have a pre-measurement stage, where sample units are prepared to be measured. And usually they have a sampling stage before the pre-measurement stage, where sample units are randomly selected from what statisticians call a population. Each stage has its sources of random error and of systematic error. The random errors in the stages compound, giving the total random error for the experiment. The systematic errors in the stages compound, giving the total systematic error for the experiment.

To keep the total random error to an acceptable minimum, researchers control the number of times the experiment is replicated, i.e., the sample size n. The larger n is, the less likely it is that $\overline{X}$ deviates much from the true value, μ. To keep the total systematic error to an acceptable minimum, researchers plan and carry out the experiment by following protocols (the details of which come later in this chapter).

2.2 Large systematic error: the invisible wrecker of experiments

Large random error visibly wrecks experiments. Large random error leaves a sign of its presence, prompting you to say: "I don't trust this experiment. Look at the wide confidence interval. It means there's too much uncertainty in the facts to draw a reliable conclusion."

Large systematic error invisibly wrecks experiments. Large systematic error leaves no sign of its presence, and that may draw you into saying, "This is a reliable experiment," when actually systematic error has made the facts greatly inaccurate. To stress how easily this can happen, consider a few notable cases. Years of wasted time passed before researchers discovered sources of systematic error in the planning and execution of the following experiments:

Systematic error was behind the experimental "discovery" of cold fusion, and behind a number of its subsequent "confirmations"(Dewdney 1997).

Systematic error was behind the "discovery" of a new form of water, dubbed polywater (Franks 1981). Experiments on its properties found its freezing point was lower than ordinary water's, and its boiling point, density, and viscosity higher. The experiments contained intolerable systematic error. It took chemists most of a decade to identify the sources. Chief among them was impurities in the experimental samples. Polywater, when the mystery was cleared up, turned out to be ordinary water contaminated with fine particulate matter.

Systematic error lies behind experimental inaccuracies in measuring fundamental physical constants of nature. I am thinking of Youden's (1972) discussion of 24 experimental estimates of the speed of light made over the years, by different researchers with different experimental designs and laboratory equipment. The researchers did their best to keep systematic error out of their experiments. It won. It got in. Of the 24 estimates of the speed of light, all were different, ranging from 91.8 to 95.1 km/sec. This spread is too wide to be accounted for by random error. That leaves the finger pointing to undetected systematic error.

Systematic error lies behind experiments confirming that a certain girl had X-ray eyes able to see medical patients' organs and diagnose their troubles, including cancer, heart disease, and kidney stones (Hyman 2005, Skolnick 2005). Later it was shown that her getting subtle clues from the patients' appearances or behaviors contaminated the experiments.

Systematic error lies behind numerous experiments that found almost perfect lie-detection accuracy in polygraph testing (Ruscio 2005). Yet when the experiments were examined carefully, systematic errors in them were found to produce a bias toward perfect accuracy.

The experiments about X-ray eyes and polygraph testing are among hundreds that have appeared in *Skeptical Inquirer* during its thirty years of publication. The magazine is devoted to debunking pseudoscientific experimental "discoveries." Frequently the debunking amounts to demonstrating how the experiments were flawed with systematic errors. (If I could, I would have *Skeptical Inquirer* be required for reading and discussing in all programs of education beyond tenth grade. This would create citizens less susceptible to believing bunk.)

When the facts of an experiment are sensational, researchers are drawn to scrutinize the plan and execution of the experiment, searching for possibly missed sources of systematic error. The trouble is, nine hundred and ninety-nine of every thousand experiments produce facts that are unsensational. With no incentive to look hard for systematic error, much of it goes unnoticed.

Even when experimenters know the sources of systematic error and minimize them, making them as small as possible, a surprising amount remains. To single out an instance in agronomy, measured soil properties are generally thought to be inaccurate by 10 to 15 percent.

The advice is clear. Plan your experiments to minimize the effects of systematic error, and carry out the plans as exactly as possible. It's difficult to do. But it can be done; witness the contents of this week's science journals.

(Besides science's dependence on data, managing all aspects of the world depends on data. Systematic error presents a risk of shifting management data away from true value, without leaving a clue it's done so. A company estimates the demand for its product, but the estimate risks being greatly biased. University administrators estimate the number of freshmen that will accept its offer of admission, but the estimate may be substantially wrong. The list is endless. Generally, systematic error in management data warps managers' views of the world, just like those warped mirrors in carnival "fun houses" – regular mirrors in which systematic error has been introduced – warp your view of yourself.)

2.3 Protocols are the key to minimizing systematic error

Protocols are procedures for accomplishing objectives. Protocols can be divided into two sorts.

One sort is protocols for designing and running machines that produce products and services. The second sort is protocols for designing and running the machines called experiments that produce facts, measurements. It's instructive to examine the fundamental difference between the two.

Consider a factory machine that produces wooden pencils. Protocols guide the machine's designers and builders. Much of the protocols comes from engineering knowledge, from knowledge of previous pencil-making machines, and from common sense. And the machine's operator is guided by a manual of protocols, do's and don'ts, as well as unwritten personal protocols formed from experience in running the machine. Notice I said, "*Much* of the protocols comes from engineering knowledge, from. . . ." The rest comes from having a target for the machine's output. The target in this case is a blueprint of the product the machine produces – a dimensioned drawing of a pencil. Look at what the machine is suppose to make (the target aimed for), and look at what the machine is making. If it's not regularly making what it's suppose to, modify the protocols for its design and its operation until it does. By the feedback, systematic error in the machine's design and operation is weeded out, as is the corresponding systematic error in the pencils produced, so that the pencils tolerably match their blueprint.

It's relatively harder to design, build, and operate an experiment that is tolerably free of systematic error. Why? Experiments are machines, but they are fundamentally different from factory machines. Whoever saw a blueprint of a measurement? The referent for telling if there is much systematic error in an experiment cannot be at the output end of the experiment.

Without a target, without a blueprint to compare the output – the measurements – to, there is no way of telling whether or not the measurements are on target. And so trial and error studies, comparing the measurements produced to a blueprint for them, are usually out. (I say usually because in rare cases it's possible to tell when the measurements produced are shifted by systematic error. If the measurements are impossibly large or small, they must be in systematic error. If a replicated experiment gives a mean value of 937 for the level of PM-2.5 in Detroit's air, that is impossibly high and obviously wrong.)

In absence of a blueprint for measurements, how do you learn protocols for designing and running experiments that can be trusted to minimize systematic error? The output end of an experiment may be in the dark, but the input end isn't. You learn trustworthy protocols in hands-on ways. You learn them by studying the manuals for operating the instruments in your field for preparing samples for measurement. You learn them by studying the manuals for operating the instruments for making measurements. You learn them from reading the "methodology section" of research reports and seeing the protocols researchers follow. You learn them from watching what other researchers do. You learn them from your experience in planning and carrying out experiments. Learning and applying them is an art.

You don't learn them from a general course devoted to minimizing systematic error. There are no such general courses. There cannot be such general courses. You've never seen a general course in minimizing systematic error, where students majoring in biology, geology, ecology, and marketing sit side by side. Protocols for minimizing systematic error differ across fields. To plan an experiment to measure a person's mitochondrial DNA, a microbiologist follows subject-specific protocols for collecting and processing the samples. To plan an experiment to measure the sidewall strength of tires, a materials scientist follows subject-specific protocols for collecting the tires, for inflating them, for varying the load on them, and for measuring the stress. Microbiologists and materials scientists know their respective field's protocols for minimizing systematic error, but neither knows the other's.

This is unlike learning the protocols for minimizing random error. They are much the same in all fields of research: students learn them in statistics courses; ecology majors and marketing majors often sit side by side.

Comment: Imagine learning to putt golf balls when you could see the target, the cup, just up until the moment you stroked the putt, but not afterwards. Unable to know how close your putts came to the target, you could not adjust your aiming and stroking to zero in on sinking your putts. You would have to trust your putting solely on protocols for lining putts up and stroking them. You and your putting experiments would be in company with scientists and their research experiments.

With the next section we are about to enter the heart of this chapter on how to plan and carry out reliable experiments – experiments with tolerable amounts of random error and systematic error. Should you feel the need to refresh your understanding of the applicable statistical terms, the Appendix explains the main ones. They are: *sample units, target population, random sample, sample size, sample mean, population mean, 95-percent confidence interval, sample size determination, nonrandom sample, standard error interval,* and *statement of accuracy.*)

2.4 The stages in planning and carrying out experiments

To be most thorough in planning and carrying out experiments, researchers must consider the experiments in terms of their stages, and focus attention on one stage at a time. The three stages are the sampling stage, the pre-measurement stage, and the measurement and calculation stage. (In medical research, the pre-measurement stage is commonly called the pre-analytic stage; the measurement and calculation stage is commonly called the analytic stage.)

Most experiments have three or two stages. Three-stage experiments consist of the sampling stage, the pre-measurement stage, and the measurement and calculation stage. Two-stage experiments have no pre-measurement stage; they consist of the sampling stage and the measurement and calculation stage.

An illustration of planning and carrying out a three-stage experiment. The following illustration features only the main considerations. It is meant to be suggestive, not final. To begin with, suppose our objective is to estimate the mean HDL cholesterol level, μ, of the workers at Hill Air Force Base. Workers are the sample units. The population of sample units is all of the workers at Hill AFB. The experiment has three stages.

The sampling stage: We plan this stage in a way that meets the protocols for sampling, as presented in books on making statistical inferences. Specifically, we determine the value of n, the sample size, by collaborating with a statistician. Let's say it comes out to be $n = 30$ workers. From the population of Hill AFB workers, we will select a random sample of 30 workers.

The pre-measurement stage: We plan this stage in a way that meets the protocols for preparing the sample units and the material gathered on them for measurement. Specifically, we will follow the blood pre-measurement protocols published by Becton, Dickinson and Company. By this, (1) we will prepare the 30 randomly selected workers by having them fast for at least 12 hours. (2) We will gather material(s) on the 30 prepared workers: we will draw a tube of blood from each, following protocol for drawing it. (3) We will prepare the gathered material(s) – the 30 tubes of drawn blood – for measuring a property of interest. Specifically, we will prepare the blood in each tube according to protocol that specifies that the blood is not to be shaken, specifies the temperature at which the blood is to be stored, prescribes how long the blood is be centrifuged, and suchlike. The result will be 30 samples of blood properly prepared for measurement.

The measurement and calculation stage: We plan this stage in a way that meets the protocols of Becton, Dickinson and Company. By this, for each of the 30 samples of prepared blood we will measure with a specified laboratory instrument the total cholesterol, x_i, where x_i indicates the measurement for the ith sample. This will give us $n = 30$ cholesterol measurements: $x_1, x_2, x_3, \ldots, x_{30}$. We will compute the sample mean of the 30 measurements, $\overline{X}$. To gauge the effect of the random error on the estimate of μ, we will compute a confidence interval. To gauge the effect of systematic error on $\overline{X}$, we will give a statement of accuracy.

Corresponding to the three stages of the plan are three stages of carrying out the plan, trying to faithfully follow the specified steps. We will not describe these actual steps. They mirror the specified steps of the plan except they are in the past tense. For instance, in place of "we will draw a tube of blood from each" the statement becomes "we drew a tube of blood from each."

Comments:

(1) Random error in the three stages combine, each contributing to the total random error in the experiment. The greater the total random error is, the greater is the variability among the n measurements we make, and so the wider is the confidence interval.

(2) Systematic error in the three stages combine, each contributing to the total systematic error in the experiment. The larger the total systematic error is, the wider the accuracy interval is, as specified by the statement of accuracy.

(3) Systematic error gets into experiments through two sources. One, the experimental design may contain sources of systematic error. On paper, in designing experiments we have to follow design protocols that will minimize the number of sources of systematic error, and

for the sources that remain we minimize the sizes of the systematic errors. Two, our execution of the design may not be faithful to the design. We may make mistakes we are unaware of making, letting systematic error slip in undetected. For, suppose the ideal (and impossible) case of perfect design protocols that promise to eliminate all possibility of systematic error. In carrying out the perfect protocols, we are at risk of systematically failing to follow parts of them. We must be careful to minimize this risk.

(4) Concerning raw measurements and derived measurements, for some kinds of research two or more raw measurements are made on each sample unit in the random sample, and/or on material prepared from them. These raw measurements are entered into calculations to produce a derived measurement for each sample unit. For instance, suppose we obtain the raw measurements of (a) total cholesterol and of (b) HDL cholesterol on each of the 30 workers. For each we divide the measured total cholesterol by the measured HDL cholesterol, giving a derived measurement, the ratio of the two. With the values of 30 ratios — $x_1, x_2, x_3, \ldots, x_{30}$ — we calculate a confidence interval on the true mean of the ratio, μ, and give a statement of accuracy for the sample mean of the ratio, $\overline{X}$.

An illustration of planning and carrying out a two-stage experiment. Two-stage experiments have a sampling stage and a measurement and calculation stage. There is no pre-measurement stage because there is no material to be gathered on the sample units and prepared for measurement. Once we have a sample of n sample units, we go directly to making measurements and calculations. All else being equal, a two-stage experiment has fewer possible sources of random error and systematic error than does a three-stage experiment.

To illustrate, consider a tract of land that has a population of $N = 4,163$ conifer trees (the sample units) on it. We want to estimate μ, the mean number of cubic feet of wood and bark per tree in the first 8-foot log of the 4,163 trees. The first 8-foot log begins 1 foot above the ground and ends at a height 9 feet above the ground.

The experiment has a sampling stage and a measurement and calculation stage. It has no pre-measurement stage because there is no need to prepare the trees for measurement. The property we are interested in is right on the trees, the amount of wood and bark in the first 8-foot log. Here is our plan for the experiment:

The sampling stage. In collaboration with a statistician, we will determine the value of n. Then from the $N = 4,163$ trees we will select a random sample of trees of the determined sample size n.

The measurement and calculation stage. For each of the n sample trees we will measure and mark two points on each tree, h_1 and h_2: h_1 is the point that is 1 foot above the ground, h_2 is the point that is 9 feet above the ground. At these two heights we will measure the tree's diameters, d_1 and d_2 respectively, using a diameter tape, an instrument of forest research made of steel to resist stretching. For each tree we will enter the four raw measurements — h_1, h_2, d_1, d_2, all in units of feet — into an equation that computes a derived measurement: the cubic feet volume of wood and bark in the first 8-foot log. We will call this "log volume," short for the volume of the first 8-foot log.

Our equation for log volume models the shape of the log as that of a parabolic frustum (this shape resembles the shape of a carton of Yoplait yogurt). Thus for the ith sample tree, the log volume, x_i, is given by the equation for the volume of a parabolic frustum:

$$x_i = ((\pi/4)(d_1)^2 + (\pi/4)(d_2)^2) \, ((h_2 - h_1))/2 \qquad (2.1)$$

With this equation we will calculate n derived measurements of log volume: $x_1, x_2, x_3, \ldots, x_n$. With the n derived measurements, we will compute the value of $\overline{X}$, the sample mean of log volume. We will estimate the confidence interval for the mean value of log volume for the population of 4,163 trees, μ. And we give a statement of accuracy for $\overline{X}$, indicating the size of the systematic error we believe is in the experiment.

So goes our plan. It remains to carry it out carefully, trying not to let any unforeseen sources of random error and systematic error accidently get in. Here's a "let's pretend" sketch of carrying it out:

According to the plan for sampling, we collaborate with a statistician and determine that the sample size should (let's say) be $n = 35$. On a computer map of the tract showing the locations of the 4,163 trees, we number them 1, 2, 3, . . . 4,163. Using random numbers, we randomly select from the $N = 4,163$ trees the $n = 35$ sample trees.

Next, we go into the forest and locate each sample tree. On each we measure h_1, h_2, d_1, and d_2. Back in our office, we enter these raw measurements into eq. 2.1 to get the $n = 35$ derived measurements of log volume: $x_1, x_2, x_3, \ldots, x_n$. To illustrate, here is how x_1, the derived log volume in tree number 1 of the 35 sample trees, is computed. For it, suppose that $h_1 = 1.0$ feet, $d_1 = 2.33$ feet, $h_2 = 9.00$ feet, and $d_2 = 2.01$ feet. Putting these measurements into eq. 2.1, we compute:

$$x_1 = ((\pi/4)(2.33)^2 + (\pi/4)(2.01)^2) \, (9 - 1)/2 = 29.75 \text{ ft}^3 \qquad (2.2)$$

Note that $x_1 = 29.75$ ft^3 is not the true value of the log volume. It is "bumped off" of true value by random errors in making the measurements, and further "slid off" from true value by systematic error. Added to this, it is further "slid off" of true value through some systematic error from treating the log's shape as a parabolic frustum, when actually its shape is slightly different and fits no regular geometric shape.

Concerning random error, even the same researcher, at approximately the same time under identical conditions, while making raw measurements might come up with different values, and thus different derived measurements. To demonstrate, imagine that we re-measure tree 1 and compute x_1 again. It might well be that this time our measurements are $h_1 = 1.0$ feet, $d_1 = 2.41$ feet, $h_2 = 9.00$ feet, $d_2 = 2.15$ feet, giving:

$$x_1 = ((\pi/4)(2.41)^2 + (\pi/4)(2.15)^2) \, (9 - 1)/2 = 32.77 \text{ ft}^3. \qquad (2.3)$$

Comparing eq. 2.2 and 2.3, the first gives 29.75 ft³ and the second gives, 32.77 ft³. If we re-measured tree 1 a hundred times, probably no two estimates of volume would be the same. Such is the effect of random error.

In any event, we measure the 35 sample trees once. For each we calculate with eq. 2.1 a derived measurement of the log volume. This gives $x_1, x_2, x_3, \ldots, x_{35}$. With these derived measurements we compute $\overline{X}$, the sample mean of the 35 trees, and compute the standard error of the mean, $S_{\overline{X}}$. Below we will show how to use $\overline{X}$ and $S_{\overline{X}}$ to compute a confidence interval for μ, the true average log volume of the $N = 4{,}163$ trees. And we will give a statement of accuracy for the size of the systematic error in $\overline{X}$.

2.5 Confidence intervals, standard error intervals, and statements of accuracy

You've run an experiment having one, two, or three stages. Each stage has some random error and some systematic error. How do you indicate the total amount of random error and the total amount of systematic error in the experiment? For random error, give a confidence interval whenever the sample is random, and give a standard error interval whenever it is not. For systematic error, give a statement of accuracy.

Below in detail, we explain confidence intervals, standard error intervals, and statements of accuracy.

How to compute and interpret a confidence interval. When your experiment involves a random sample of facts from a population, and when you replicate it enough times to give a confidence interval that should be narrow enough to be meaningful, then report a confidence interval.

The equation for a 95-percent confidence interval (contained in introductory texts on inferential statistics) is:

$$\overline{X} \pm (S_{\overline{X}})(t), \text{ sometimes written as: } \overline{X} - (S_{\overline{X}})(t) \leq \mu \leq \overline{X} + (S_{\overline{X}})(t). \quad (2.4)$$

Here, $\overline{X}$ is the sample mean; $S_{\overline{X}}$ is the sample standard error of the mean; t is the t-value for 95-percent confidence. (For other-percent confidence intervals, as 90-percent, the value of t in eq. 2.4 is for that percent confidence.)

To demonstrate the calculation of a 95-percent confidence interval, we continue with the tree example of section 2.4. Recall we had $n = 35$ derived measurements of the log volume: $x_1, x_2, x_3, \ldots, x_{35}$. With the 35, we computed the sample mean, $\overline{X}$, i.e., the mean log volume. Let's say we got $\overline{X} = 30.9$ ft³. And we computed the estimated standard error of the mean, $S_{\overline{X}}$. Let's say we got $S_{\overline{X}} = 1.9$ ft³. We then look up the value of t in a t-table for 95-percent confidence; it is 1.96. Putting these values of $\overline{X}$, $S_{\overline{X}}$, and t into eq. 2.4 gives the 95-percent confidence interval estimate for the mean log volume, μ, of the population of 4,163 trees:

$$30.9 \pm (1.9)(1.96), \text{ which is:}$$
$$30.9 \pm 3.7 \text{ ft}^3, \text{ or } 27.2 \leq \mu \leq 34.6 \text{ ft}^3. \qquad (2.5)$$

Here is what it means. If we drew a second random sample of size 35 trees, measured them and computed the 35 estimates of log volume, we could compute a second 95-percent confidence interval. If we took a third random sample of size 35 trees, measured them and computed the 35 estimates of log volumes, we could compute a third 95-percent confidence interval. Suppose we went on and on, a million times. We would get a million 95-percent confidence intervals, each an estimate of the interval containing the mean log volume, μ, of the population of the 4,163 trees.

Because these are 95-percent confidence intervals, out of 1,000,000 of them, 950,000 will contain the value of μ in their interval. The other 50,000 won't; they are lies, implying they contain the value of μ when they don't. (Actually it might be something like 950,168 will contain μ in their interval. To get exactly 95 percent requires sampling an infinite number of times, not 1,000,000 times. This has no consequences for our discussion here.)

Look again at eq. 2.5. It's our 95-percent confidence interval, the one we got. Is it a lie? Well, there is a 95-percent chance that it is one of the million that is truthful. It is in this sense that there is a 95-percent chance that the mean log volume, μ, of the population of 4,163 trees lies somewhere between 27.2 ft^3 and 34.6 ft^3. In saying this, we are assuming that we planned and carried out the experiment on the trees in such a way that it contains negligible systematic error. Appreciable systematic error would appreciably make $\overline{X}$, as well as the confidence interval, be lies. Three additional points:

(1) The greater the random error in our experiment, the greater would be the average width of the million confidence intervals.

(2) Confidence intervals reflect the total random error in experiments. In our experiment to estimate the mean log volume of the $N = 4,163$ trees, there was random error in sampling the trees, and random error in measuring the trees. Equation 2.5 gauges the combined effects of the two, the total random error.

(3) In our experiment, we have underestimated the total random error. It really is larger than our confidence interval accounts for. This is because (as frequently happens with experiments) we haven't considered all of the sources of random error. In particular, we assumed we measured each tree exactly at the marks $h_1 = 1.0$ feet and $h_2 = 9.0$ feet, and so we put these values into eq 2.1. In actuality, on each tree we were randomly a bit off in locating h_1 at 1.0 feet and h_2 at 9.0 feet. The effects of these random errors are not accounted for in our analysis.

How to compute and interpret a standard error interval. To compute a standard error interval, compute the sample mean, $\overline{X}$, and the standard error of the mean, $s_{\overline{X}}$, and combine them in the equation for the standard error interval, viz.:

$$\overline{X} \pm s_{\overline{X}} \text{ (or if you prefer, } \overline{X} \pm \text{SE}). \qquad (2.6)$$

There are three instances when you should report a standard error interval instead of a confidence interval:

(1) Report a standard error interval whenever you take a nonrandom sample from a population. To demonstrate, suppose you want to measure the dust deposition rate at a site. For this you install a grid of dust collectors. The sample of dust that the wind haphazardly brings to the collectors is scarcely a random sample. As such, don't compute and report a confidence interval; it would be invalid. You should compute and report a standard error interval. It has validity for random samples and non-random samples.

(2) Report a standard error interval whenever the sample isn't a sample from a population but actually is the entire population. Here's an example. An archaeologist enters a cave and finds six beads on the floor, the only six in the cave. The archaeologist computes their mean weight and standard error. It makes no sense to compute a confidence interval on the true mean weight because the six comprise the entire population. Therefore, the archaeologist reports a standard error interval.

(3) Report a standard error interval when standard error intervals are your preferred measure. Some researchers, even when they work with random samples, and even when the samples are large enough to produce narrow confidence intervals, simply prefer to think in terms of standard error intervals. The following research example illustrates the preference.

Res. Ex. 2.1: Estimating the incubation time of a disease

Angers et al. (2006) inoculated eight mice with extracts of prions prepared from muscle tissue of mule deer suffering from Chronic Wasting Disease (CWD), and they inoculated six mice with extracts of prions prepared from brain tissue of mule deer suffering from CWD. On each mouse in each of the two groups, Angers et al. measured the incubation time, i.e., the number of days from the day of inoculation until the mice developed prion disease.

They reported a standard error interval for each group. For the group inoculated with prions from muscle tissue, it was 367 ± 9 days. For the group inoculated with brain tissue, it was 278 ± 11 days. These standard error intervals provide readers of the report with the mean incubation time for prion disease, and with the variability in the time.

These and other of Angers et al.'s research findings prompted their conclusion that "humans consuming or handling meat from CWD-infected deer are therefore at risk to prion exposure." Even though they did not strictly select random samples of mice, most people would consider their conclusion to be sound. (This research is an example of cause-effect research, planned and carried out with what is called "the direct method," as explained in chapter 5.)

How to determine a statement of accuracy. A **statement of accuracy** is a subjective estimate of the amount of systematic error in an experiment. After you have carefully and thoroughly done your best to make sure that your experiment is free of all systematic error, some amount of systematic error will remain. There may be sources of systematic error present you don't know about. And there will surely be sources of systematic error you minimized but could not eliminate. The remaining systematic error causes inaccuracy in the experimental facts – x_1, x_2, x_3, . . . , x_n – sliding the set of them off of the true value by a certain amount. A statement of accuracy informs researchers of the magnitude of the sliding, not the direction.

Eisenhart (1968) recommends that researchers report a statement of accuracy. It indicates the largest amount of systematic error they believe could be in their experiment. He recommends phrasing the statement of accuracy along any of these lines:

(1) ". . . is (are) not in error by more than 1 part in (y)."

(2) ". . . is (are) accurate within ± (i) units [or ± (i) percent]. "

(3) ". . . is (are) believed accurate within (. . . .)."

These are statements of accuracy. For the given research, choose the most fitting form.

To demonstrate, consider the second form. The symbol i represents the most the inaccuracy is expected to be. And consider the above tree experiment of estimating the mean log volume for the sample of $n = 35$ trees. Suppose we decide that i = 1.5 ft^3. This means we believe that the mean volume of the 35 trees, $\overline{X}$, is inaccurate by no more than ± 1.5 ft^3, due to systematic error in the experiment. Recalling that $\overline{X}$ = 30.9 ft^3, in the report of the experiment we would write a statement of accuracy like this: "Considering the systematic error in the experiment, the sample mean, 30.9 ft^3, is believed accurate within ± 1.5 ft^3." This means we would be very surprised if the systematic error in the experiment caused the mean volume of the 35 trees to be outside the range from 29.4 ft^3 to 32.4 ft^3.

Another of Eisenhart's forms for a statement of accuracy expresses the bounds of accuracy as a percentage, p, of the mean, $\overline{X}$, viz. $\overline{X}$ ± p percent, which reads: "The sample mean, $\overline{X}$, is believed accurate within ± p percent"; e.g., 10 percent.

Subjectively estimating i (or p). In general, base your estimate of i (or p) on your knowledge and intuitive sense of the systematic error in your experiment, as well as on systematic errors in any similar experiments you are familiar with. In particular, go through the following steps. Begin with a value for i of 0.0 (or for p of 0.0), and ask yourself if that is likely to be realistic. You will say "no" because you know that some systematic error must be present. Increase i by 5 percent, say, and ask yourself if that is likely to be realistic. If you say no, keep increasing i until you reach a value such that if you made it any larger you would suspect it would be less likely to be realistic. This is the proper value for the statement of accuracy. Any larger value makes you doubt it is that large; any smaller makes you doubt it is that small.

How much inaccuracy is too much? Let I* (or P*) be the largest value of i (or p) that you consider acceptable. Choose the value of I* before you do the experiment, and hold to it. By no means let the value of I* affect your deciding the value of i. If the comparison of i with I* comes out that i ≤ I*, conclude that there was a tolerable amount of systematic error in the experiment. But if i exceeds I*, declare the experiment is a failure, meaning the estimate $\overline{X}$ is intolerably unreliable because of systematic error. Absolutely resist the temptation to fudge i or I*, or both, for the sake of declaring that the experiment is reliable. This is what an ideal researcher would do, though I concede such researchers are in short supply. It's time to change that.

Information you should include in your research report. In your research report describing your experiment: (1) Put the sources of systematic error that you considered, and say how you minimized their effects. (2) Put the statement of accuracy, giving the value of i (or of p). (3) Give the value of I* (or P*). These three items will help readers understand and evaluate your decisions regarding systematic error.

More than ninety-nine of every one hundred researchers report none of this information. Don't do as they do. For the sake of increasing the amount of reliable knowledge in the world, do as this book says.

2.6 Some illustrative sources of systematic errors

The intent of this section is threefold. First, I would like to increase your sense of some common kinds of systematic error, and put you on guard to prevent them from occurring in your experiments. Second, I would like to impress you with the idea of how easy it is to overlook systematic errors, letting them slip into your experiments. Third, I would like you to carry your increased awareness of systematic error into critiquing the research papers and seminars of researchers.

Below are illustrations of some kinds of systematic errors that can get into the sampling stage, into the pre-measurement stage, and into the measurement and calculation stage. The illustrations are a mix from experiments in the social sciences, the life sciences, and the physical sciences.

In the physical sciences, researchers tend to prefer the term "systematic error." In the social sciences, researchers tend to prefer the term "bias." Both terms mean the same. Each illustration below adopts the term that researchers in the context of the illustration seem to favor.

Illustrative sources of systematic errors in the sampling stage. *Sampling a non-target population.* Mary, a medical researcher, defines the target population to be all citizens of Nevada over the age of 69 who in the last five years survived a heart attack. She selects a random sample of them and sends them a questionnaire about the state of their health. The trouble is that those in poorer health, possibly because of heart problems, are less likely to respond to the questionnaire. Such non-response could bias the estimate of the mean state of health of the target population.

Sampling a non-target population. As school principal, Joe defines the target population to be the school's ninth graders on a typical school day. He plans to have teachers measure the ninth graders' I.Q.'s with the Sanford-Benet I.Q. test. It will be a 100-percent sample, a complete census. But there is a contingency Joe fails to notice, and it produces a systematic error. It is influenza season; some of the children stay home on the day of the test, and some who come don't feel well. The population in school that day is atypical, a non-target population. The mean I.Q. score of that population is likely biased from the mean I.Q. score of the target population.

Sampling a non-target population. As a medical researcher, Alison defines the target population to be patients diagnosed with multiple sclerosis (MS). But she mistakenly samples a non-target population. She has selected a random sample of patients for a clinical trial to test a novel drug for treating MS. But not all patients in the sample belong to the target population. Some have been misdiagnosed and do not have MS. Rather they have diseases with symptoms similar to those of MS, such as Lyme disease, Parkinson's disease, and juvenile rheumatoid arthritis. The systematic error leads Alison to err in estimating the drug's effect on MS patients.

Violating one or more assumptions of inferential statistics. Collin, a forester, plans to select a random sample of trees from an uneven-aged stand of 630 Subalpine fir. With an increment borer, he'll extract a core from each sample tree, and count the rings as a measure of the tree's age. Using the measured ages, he'll compute a 95-percent confidence interval on the average age of the stand. Now, judging from the heights of the trees, the frequency distribution of the 630 ages is not a normal distribution, with many young trees and fewer old ones. Because of this, an approximate rule for making eq. 2.4 give an unbiased estimate of the 95-percent confidence interval is to have the size of the random sample be at least $n = 25$ (Barrett and Nutt 1979). But Collin doesn't know the rule or it slips his mind, and he goes ahead and randomly selects fewer trees, $n = 15$. With the 15 ages, he computes a 95-percent confidence interval. But it is wrong. Because the sample size is smaller than required, the percent confidence on the interval is less than 95 percent, by some unknown percent. It's in systematic error.

In general, failing to meet one or more assumptions introduces systematic error into confidence intervals and tests of statistical hypotheses.

Illustrative sources of systematic errors in the pre-measurement stage.

Incorrectly preparing the sample material. Consider Patsy, a plant scientist. The protocol for her experiment calls for preparing samples of grasses for measurement by drying them in a forced-air drying oven at 60°C for a prescribed period. Patsy commits a systematic error when she mistakenly sets the temperature at 50°C. Later, in weighing the samples, she's unaware that the weights are systematically greater than they would be had the protocol been followed.

Contaminating the sample material. Jeff, an environmental chemist, is researching certain effects of acid rain. The protocol calls for random samples of lake water to be collected. This he does; however, he doesn't notice that he accidently contaminates the samples with alkaline matter. Later, applying the method of titration to determine the acidity of the samples, unbeknown to him the results are systematically in error.

Contaminating the sample material. Laurie, an archeologist, takes five random samples of bits of wood charcoal from an ancient site on a Polynesian island. But she takes no precautions to insure that the bits are free of carbon from wood or marine organisms from an earlier or later age. Unbeknown to her, the samples are contaminated with carbonous material that is older than the wood charcoal. This causes a systematic error in the radiocarbon dates of the five samples, making them appear older than they are.

Inadvertently modifying the sample material. Kim's research calls for her to estimate the bulk density of soil in a natural area. However, she fails to realize when she drove her truck in last night in the dark she drove it over the area she is going to sample, compacting the soil, which artificially increases the bulk density.

Illustrative sources of systematic errors in the measurement and calculation

stage. *Failing to calibrate the measuring instrument.* Jacob, a nuclear engineer, uses a neutron probe to measure the neutron flux at various positions in a sub-critical fission assembly. But despite his careful intentions, he commits a systematic error by failing to calibrate the probe against a known neutron flux. Consequently he's unaware that the flux values he measures differ from the true values.

Failing to insulate the measuring instrument from conditions that could make it malfunction. While counting beta particles with a liquid scintillation counter, Charles unwittingly lets a systematic error occur. He's not aware that whenever people upstairs turn on the lights, it causes electrical line noise, which the scintillation counter records as counts of beta radiation. (Similar situations include operating measuring instruments in harsh environments they aren't designed for, as in extreme cold, heat, or dampness.)

Failing to use the measuring instrument as its manual prescribes. Stephen, a materials researcher, does an experiment to determine the density of a liquid aluminum alloy. He pours the metal into a cylindrical pycnometer, which has a known volume stated by its manufacturer. According to the manual, a meniscus may form around the pycnometer's top internal corners, making the poured volume be less than the pycnometer's stated volume. This happens in Stephen's experiment. When he empties the pycnometer, weighs the metal, and computes its density by dividing its weight by the pycnometer's stated volume, the density he calculates is less than the true density. Although he replicates the experiment four times, each time the computed mean density is systematically less than its true density. Not having closely read the pychometer's manual, Stephen failed to consider a table in it with correction factors for the meniscus effect for various liquid metals, including aluminum. (In general, to not know an instrument's manual forwards and backwards is to risk making systematic errors with the instrument.)

Allowing observer-expectancy bias to affect measurements. This can happen in experiments where an observer is the measuring instrument. Observer-expectancy bias is the tendency for an observer to subjectively judge experimental outcomes in the direction of the observer's expectation, thus making a systematic error.

As a researcher in animal behavior, Dalin tests a research hypothesis specifying which of the grizzly bears feeding on salmon at McNeil River Falls, Alaska, will dominate other bears, according to their sexes, sizes, and ages. While observing the bears' behaviors and judging dominance, Dalin unconsciously allows his expectation from the hypothesis to influence his judgments in the direction of corroborating the hypothesis.

Doing a single-blind experiment, where observers have no expectations about the meanings of the possible outcomes, can minimize observer-expectancy bias. (See Balph and Balph (1983) for more on the problem of observer-expectancy bias in animal behavior experiments; the article quotes a researcher who summed up the problem with the line, "I wouldn't have seen it if I hadn't believed it.")

Allowing subject-expectancy bias to affect measurements. Also known as the placebo effect, subject-expectancy bias is the tendency of human subjects to respond as they believe the researcher observing them expects them to, or as they themselves expect to.

As a medical researcher, Rachel possibly commits a systematic error in an experiment to estimate the average degree that acupuncture relieves wrist pain thought to be caused by sleeping on one's hands. With acupuncture she treats a random sample of patients who sleep on their hands and have the pain. Afterwards she asks them to rate on a 5-point scale the degree of relief they experienced, relative to the level of pain before the treatment. There is a positive degree of relief. The trouble is that it may have been caused by the patients' expectation that the treatment would work.

A related point: In medicine research, double-blind experiments minimize observer-expectancy bias and subject-expectancy bias. With a double-blind experiment, researchers don't know which patients receive the treatment and which receive a placebo, nor do the patients know what they received. Yet even double-blind experiments require careful attention in their design, for systematic errors can be made by imperfectly "blinding" the patients receiving treatments and/or the researchers evaluating treatments (Jadad 1998).

Letting personal sympathies influence subjective ratings. In her social work, Sara administers an attention-deficit-syndrome test to a random sample of hyperactive children. The test asks her to judge the children on a number of behaviors, such as inattention in listening and completing tasks, restlessness, and impulsively acting without thinking. But unconsciously she commits a systematic error. She is more sympathetic to some of the children – her favorites – than to others. This clouds her professional judgments. Her ratings of her favorites are higher than they would be if she were not especially fond of them.

Allowing the act of measuring to affect the measurements. As a clinical nurse, Kristopher measures the blood pressures of a random sample of elderly people. But the measurements are systematically elevated because having their blood pressure measured makes the patients anxious. Kristopher failed to follow standard protocol to calm the patients minutes before taking the blood pressure measurements.

Allowing the act of measuring to affect the measurements. Mylie, a wildlife researcher, equips cormorants with radio responders, allowing her to measure their locations for mapping out their home ranges. However, the responders cause them to modify their travel habits, hence modifying their home ranges.

Using a biased database. Sean, a sociologist, wants estimates of the poverty rates of single mothers in certain U.S. cities. He goes to a database touted as containing the percent of single mothers below the official U.S.-defined poverty line. In using this database, Sean commits a systematic error because the compilers of the data included only the wages of single mothers. To the point of this, Jencks (2006) says: "The official poverty rates for single mothers. . . . have numerous problems: they overestimate inflation since the poverty line was established; they ignore noncash benefits like food stamps and housing subsidies; they ignore taxes and the Earned Income Tax Credit; and they ignore the income of men who live in a mother's household but are not married to her." (Jencks goes on to cite a researcher who developed an adjusted measure, reducing most of the systematic error.)

Using an incorrectly defined measure. Cecilia's a paleontologist. Her experiment calls for her to estimate the average volume of a sample of fossilized spherical pollen grains. She measures the diameters of the grains, from the measurements computes the average diameter, and from the average diameter she computes the average volume. This measure of average volume is in systematic error. The correct calculation, which is free of systematic error, goes this way. Take the measured diameters of each grain, compute the grain's volume, and average the volumes of all the grains. Looked at another way, the volume of a pollen grain that has the average diameter (that's what she computed) is systematically less than the average of the pollen grains' volumes (which her research calls for).

Using spuriously correlated measures. Spuriously correlated measures are systematically in error. McCuen (1974) describes a case in outdoor recreation research. The researcher (let's call him Justin) selects a random sample of 20 recreation sites having lakes. For each site he measures three variables: the number of people who visited the site in July (V), the number of people living in the surrounding area (P), and the lake surface area in acres (A). From these raw measurements, he derives two ratios: $Y = A/P$ (acres of lake per population), and $X = V/P$ (visitations per population). Then Justin unknowingly makes a systematic error. Plotting Y versus X, he sees an impressive correlation between the two. Mathematically he gauges its strength: he computes the correlation coefficient between the two; it is 0.72 and statistically significant. Yet as McCuen (1974) demonstrates with the data, the correlation between V and P is nil, the correlation between V and A is nil, and the correlation between A and P is nil. Further, the correlation between Y and X is actually nil: the value of 0.72 is illusory, a case of spurious correlation due to Y and X containing the common variable P.

Consider another example of spurious correlation. Eberhardt (1970) cites a case of ecologists who measured N_t and N_{t+1}, the numbers of a certain animal in an area at date t and later at date t+1. With the data, they computed the value of the correlation coefficient between the variable N_t and a derived variable, the ratio N_{t+1}/N_t. The value was impressive, about -0.7, but they didn't realize it was spurious. Later, Eberhardt showed mathematically that the correlation of N_t with N_{t+1}/N_t, when N_t and N_{t+1} are random numbers, is -0.707. The reason is that N_t is common to the two variables (N_t, N_{t+1}/N_t), forcing a spurious correlation.

Cases of spurious correlation in hydraulics and hydrology are reported by Benson (1965), and in forestry by Dean and Cao (2003). Cases in some other fields are presented by Butler (1986), Garsd (1984), Jackson and Somers (1991), Kenney (1982), Kite (1989), Phillips (1986), Scott (1979), and Sockloff (1976). A clearly written introductory book on spurious correlation, with examples from geology, is Chayes' (1971) *Ratio Correlation*.

Using a wrong baseline for comparison. Holly, an environmental health researcher, analyzes the death records of 22,000 people who worked around hazardous chemicals at a microprocessor manufacturing company during the past 20 years. From the data she calculates the cancer mortality and cancer incidence rates. For the baseline rate she chooses the rate in the general population. She finds that the rate for the company is nearly the same as the baseline rate. The trouble is, this is the wrong baseline rate. Employed people are healthier than unemployed people, and the general population contains a good proportion of unemployed people. In reality, the rates for the microprocessor company may be higher than the proper baseline rate: the rate for workers in companies who weren't exposed to chemicals.

Comment: Time out for a test. Raloff (2008) quotes a Harvard University scientist saying that experiments should be "replicated to ensure reliability." Is it true that replication ensures reliability? I hope you are thinking: "Not necessarily." Replication ensures the reliability of experimental facts only when the systematic error in the facts is negligibly small.

When you replicate your original experiment, the replication follows the same plan and uses the same equipment that the original experiment did. Whatever systematic error is in the original experiment is in the replicated experiment. Even though the original and replicated means may closely agree, it is a close agreement of erroneous, inaccurate means.

2.7 The pre-measurement stage is potentially thick with sources of systematic error

The intent of this section is to make a big lasting impression on you that often the pre-measurement stage contains dozens of potential sources of systematic error. So let's look at a hypothetical but realistic experiment for measuring certain blood chemistry analytical values of people. We'll list the systematic errors that, unless minimized, will spoil the preparation of blood samples in the pre-measurement stage. For minimizing the systematic errors we will give a clinical laboratory protocol established by Becton, Dickinson and Company.

Because your field is probably unrelated to measuring blood chemistry analytical values, this section may induce intense yawning. My advice is to bear through it for the value of the big impression it makes, which will carry through to your field of interest. I want to drive it into your bones that many sources of systematic errors lurk in the pre-measurement stage of experiments in every field. Knowing this is reason to rigorously search for them, put them out in the open to address them, doing whatever is necessary to minimize them.

Listed below for the pre-measurement stage are twenty-three potential sources of systematic errors and the protocols for minimizing them. They are arranged under three categories: specimen collection issues; processing, handling, and transport issues; and physiological issues.

SPECIMEN COLLECTION ISSUES
1. *Systematic error:* Leaving the tourniquet on the patient's arm for an extended period of time risks causing hemoconcentration and possible hematoma due to the infiltration of plasma and/or blood into tissue. This can affect the water balance of cells. It can cause red cells and platelets to rupture and release potassium. *Protocol:* Release the tourniquet as soon as blood flow is established; tourniquet should be released within 1 minute.
2. *Systematic error:* Repeated fist clenching with or without tourniquet can result in excessive release of potassium from skeletal muscles (pseudohyperkalemia). *Protocol:* Ask patient to dangle the arm for 1 to 2 minutes to allow blood to fill the veins to capacity; then reapply the tourniquet. Massage the arm from wrist to elbow. Tap sharply at the venipuncture site with the index and second finger a few times. This will cause the vein to dilate. Apply

a warm, damp washcloth (about 40°C) to the site for 5 minutes. Avoid fist clenching during phlebotomy.

3. *Systematic error:* Having the arm in an upward position can cause reflux or "backflow" from anticoagulants (EDTA, Oxalate/Fluoride), and carryover from the previous tube. *Protocol:* Position arm downward. Follow recommended order of drawing blood.

4. *Systematic error:* Having Betadine (antiseptic) on the arm. When samples are selected at the same time as starting an IV, Betadine can cause an increase in potassium results. *Protocol:* Completely remove Betadine using 70% alcohol prior to venipuncture. Use a discard tube to remove the first few mLs of blood.

5. *Systematic error:* Drawing blood into lavender-top potassium EDTA tubes before drawing into serum chemistry tubes, and drawing into gray-top potassium oxalate/sodium fluoride tubes before drawing into serum chemistry tubes. This can cause a carryover of potassium-containing anticoagulants into serum tubes. *Protocol:* Draw serum and heparin tubes prior to lavender- or gray-top tubes during the collection procedure. Follow the recommended order of draw: 1. Blood culture tubes. 2. Non-additive tubes. 3. Additive tubes.

6. *Systematic error:* Drawing above IV site. This can cause IV fluid contamination. *Protocol:* Draw below IV site or at alternate site.

7. *Systematic error:* Using catheters coated with benzalkonium heparin. This can cause interferences and falsely high reading with some ion-selective electrodes. *Protocol:* Clear the catheter line by withdrawing and discarding 5 mL of blood. This procedure is not sufficient if blood is drawn through a newly inserted catheter; in that case, collect specimen via direct venipuncture.

8. *Systematic error:* Vigorously mixing tubes. This can cause hemolysis due to rupture of red blood cells. *Protocol:* Gently mix additive tube using the recommended number of inversions.

9. *Systematic error:* Having faulty collection technique involving small gauge needles, syringe/catheter draws, or transfer of blood into evacuated tubes. Such can cause hemolysis. *Protocol:* Pay attention to correct technique. Use partial draw tubes to minimize turbulence. Use BD Vacutainer® Blood Transfer Device to move blood from syringe into evacuated tube.

10. *Systematic error:* Making a traumatic draw. This can cause hemolysis. *Protocol:* Select appropriate vein size for required volume of blood. Do not probe.

PROCESSING, HANDLING, AND TRANSPORT ISSUES

11. *Systematic error:* Using pneumatic tube systems that excessively apply forces to moving canisters, namely systems that accelerate the canisters too rapidly, or employing unpadded canisters or stations, causing excessive agitation of transported samples. Such can cause red blood cell trauma and damage. *Protocol:* Use low speed pneumatic tube system. Adequately pack the specimens to avoid excessive mixing.

12. *Systematic error:* Allowing delays in processing/transport can cause release of potassium from cells. *Protocol:* Serum/plasma should be removed/separated from cells within 2 hours of collection.

13. *Systematic error:* Centrifugation with g-force set too high, with undue heat exposure during centrifuging, or with running fixed-angle centrifuge continuously for long periods of time can cause lysis of cells. *Protocol:* Observe the following: 1000-1300 g-force for BD Vacutainer® SST™ Glass Tubes and 16 mm BD Vacutainer® SST™ Plus Plastic Tubes; 1100-1300 g-force for BD Vacutainer® SST ™ Plus Plastic Tubes (13mm); < 1300 g-force for all non-gel tubes. Use temperature-regulated centrifuge, and follow manufacturer's recommendation for centrifugation time and g-force.

14. *Systematic error:* Re-centrifugation can cause mixing of serum below the gel with serum above the gel. *Protocol:* Do not re-centrifuge BD Vacutainer® SST™ Tubes. Aspirate serum from tube and place in a clean test tube to re-centrifuge.

15. *Systematic error:* Poor barrier formation in gel tubes. This can cause red blood cells to get above gel, leakage of RBCs across barrier, RBC contamination, falsely high potassiums, and other erroneous test results. *Protocol:* Follow manufacturer's recommendation for centrifugation time and g-force. Invert BD Vacutainer® SST™ Tubes gently 5 times immediately after specimen collection. Allow tube to clot 30 minutes in a vertical position. Centrifuge samples for 10 minutes in a horizontal swing bucket, 15 minutes in a fixed angle. Apply 1000-1300 g-force for BD Vacutainer® SST™ Glass Tubes and 16 mm BD Vacutainer® SST™ Plus Plastic Tubes; 1100-1300 g-force for BD Vacutainer® SST™Plus Plastic Tubes (13 mm); <1300 g-force for all non-gel tubes. Periodically check/calibrate centrifuges. Use swing bucket centrifuge. Do not re-centrifuge gel tubes. Transfer serum to another tube if re-spinning is necessary.

16. *Systematic error:* Chilling whole blood beyond 2 hours risks inhibiting glycolysis which provides energy for pumping potassium into the cell. Without this energy, potassium will leak from the cells, falsely elevating the results. *Protocol:* To chill a sample, place it either in crushed ice or a mixture of ice and water, for less than 2 hours. Do not chill < 15°C.

PHYSIOLOGICAL ISSUES

17. *Systematic error:* Thrombocytosis, and/or myeloproliferative disorders with severe leukocytosis risk causing platelets to release potassium during clotting in serum. An increase of 1 million platelets/μL corresponds to an increase of about 0.7 mEQ/L in the serum potassium; in plasma more platelets remain above the gel barrier. *Protocol:* Allow complete clot formation. Centrifuge at the high end of recommended centrifugation range.

18. *Systematic error:* Dehydration can cause inherent higher potassium levels, depending on patient's condition. *Protocol:* Hydrate patient and then re-draw specimen.

19. *Systematic error:* Anticoagulant therapy (Coumadin, Heparin), and/or liver disease can cause medically induced delays in the clotting process. If tube is re-spun, serum below barrier (higher potassium) mixes with serum above barrier. *Protocol:* Note that 30-minute

clotting may not be sufficient. Observe clot formation up to 1 hour. Transfer serum to another tube if re-spinning is necessary. Use heparinized plasma for potassium analysis. Add thrombin to accelerate clotting.

20. *Systematic error:* Patient's fear of imminent venipuncture can lead to acute hyperventilation and a net potassium efflux from cells. *Protocol:* Ease patient's fears about the procedure.

21. *Systematic error:* Hereditary pseudohyperkalemia risks an abnormal passive leak of potassium across the RBC membrane especially at lower temperatures, because of autosomal dominant loci on chromosome. *Protocol:* Check patient's familial history. If pseudohyperkalemia present and potassium is out of range when measured, pseudohyperkalemia is possibly the reason.

22. *Systematic error:* Oral therapy of Cotrimoxazole can lead to hyperkalemia with renal tubular dysfunction. *Protocol:* Discontinue Cotrimoxazole to normalize serum potassium levels and symptoms.

23. *Systematic error:* Serum vs. Plasma: potassium is greater in serum than in plasma due to release of K+ from platelets during clotting. Plasma potassium increases over time due to presence of cells in plasma. *Protocol:* Standardize on either specimen type to establish normal reporting ranges for both serum and plasma. Centrifuge within 2 hours. Aspirate plasma from tube, put in clean tube, and re-spin plasma.

Whatever the field, the pre-measurement stage of experiments is a potential minefield of systematic error. So much can go wrong there. Be extra careful and thorough to locate every pre-measurement source of systematic error in order to minimize it.

2.8 The advantages of standard constructs

If you were explaining to a lay person what a construct is, you would say something like, "A construct is the thing you measure." If explaining it to a researcher who didn't know, you would do well to take the definition of a construct given by Caws (1957) and translate it to this simple and direct language: "A **construct** is a constructed concept – hence the name – that can be objectively measured. Constructs are constructed by giving them operational definitions – definitions in terms of the operations required to measure the constructs. Inherent in the operational definition of a construct is the construct's meaning and the protocol for measuring it."

What we are calling a "standard construct" is what Caws simply calls a "construct." This allows us to distinguish a "nonstandard construct" (more about it later).

Standard constructs prescribe the operations an experimenter is to take in making measurements, and prescribe any special steps for preparing the material to be measured. Thus in keeping systematic error at bay, standard constructs encompass the measurement and calculation stage, and the pre-measurement stage, of experiments. Their scope does not extend to the sampling stage.

The advantages of using standard constructs are objectivity, unbiasedness, and comparability (Caws 1957). Objectivity: standard constructs are operationally defined. Unbiasedness: by their definitions there is no systematic error in measuring them. Comparability: they provide a common basis for comparing the facts of different experiments.

Research societies, by the research journals they support, promote the use of standard constructs. By reading the research journals in your field, over time you learn of the standard constructs, which researchers in your field regularly use in their experiments. To illustrate:

The Sanford-Benet I.Q. test is a standard construct for measuring intelligence. Another is the Otis self-administering test of mental ability.

The Allport-Vernon test scale of values is a standard construct for measuring people's values. With it, a person ranks his or her preferences on a series of test items, such as "A good government should aim chiefly at: (A) more aid for the poor, sick, and old; (B) the development of manufacturing and trade; (C) introducing the highest ethical principles into its policies; (D) establishing a position of power and respect among nations." After the person answers the test items, certain economic values, religious values, and social values can be scored on a point scale. In this way, a population of people can be sampled and their values measured objectively in a widely accepted way.

The Rockwell hardness test is a standard construct for measuring the hardness of metallic materials or plastic materials. The test spells out how a sample of material is to be struck with a standardized indenter at a prescribed force, and how the degree of indentation is to be scored, this giving the measurement of the material's hardness. Besides Rockwell hardness, other standard constructs for measuring hardness are the Brinell hardness test and the Vickers hardness test.

In general, an instrument and the manual prescribing how to operate it, and, if relevant how to prepare the material it is to be operated on, define a standard construct. Examples include an electronic thermometer to measure temperature, an electronic hygrometer to measure humidity, an electronic scale to measure weight, a spectrophotometer to measure genetic structure from DNA, and a Geiger counter to measure ionizing radiation.

To minimize systematic error in the measurements, researchers calibrate the instruments against known standards, or periodically return them to the manufacturers to have them calibrated.

The history of science shows that certain nonstandard constructs gain acceptance over time and scientists widely come to regard them as standard. I'll tell you one field where I expect this to happen (it is not the only one). It's wildlife science. Wildlife science has no standard construct for the concept of "carrying capacity," a measure of the capacity of habitat to continually support populations of large mammals, such as deer. Rather, wildlife science has more than a dozen nonstandard constructs of carrying capacity. In time the list will probably be winnowed down to one or two constructs, and they will be regarded as standard. Winnowing down occurs when one construct is shown to serve better for theories, laws, and practical purposes (for more, see Romesburg 1981).

2.9 Systematic errors don't always cancel out in comparative experiments

In the simplest case, a comparative experiment involves two experiments producing facts that are compared. Types include before-and-after experiments, and treatment-and-control experiments. Another type has no formal name; it is two experiments that measure a quantity at two times or two places. An example is two experiments that measure the thickness of the Arctic ice pack at a certain place in the Northwest Passage. One experiment is run one year, the other the next. Comparing the mean measurements from the two, the research answers: "Has the thickness of the ice pack changed?" and, "If it has, what is the change?" Another example is two marketing experiments to measure consumer demand for a certain product in two regions of a country. Comparing the mean demands from the two, the research answers: "Are the mean demands different or the same?" and, "If different, what is the difference?"

More complex cases have multiple experiments. Rather then one treatment and a control, there are multiple treatments, with the corresponding measurements from each treatment experiment compared with each other and with measurements from the control experiment. Or there may be a series of experiments at a series of times, with measurements from any pair of experiments in the series compared, this producing a graph of mean measurements against time (such as mean measurements of the Arctic ice pack made yearly for a period of years).

On occasion I've heard researchers claim that systematic errors are "no big deal" in comparative experiments because they cancel out in the comparison. The claim, as a blanket statement, is false. Sometimes they cancel, sometimes they don't. Mainly we will illustrate this with the simplest case of a comparative experiment, i.e., two experiments. What holds for this holds for complex cases, i.e., multiple experiments. (For situations illustrated below, we will assume there is no random error in the experiments. This makes for clear explanations without having any effect on the conclusions we'll draw about systematic error.)

A kind of situation where systematic errors cancel out. When the raw measurements in the two experiments are shifted from their true values by the same systematic error, the difference in the raw measurements and in their means is unaffected by the systematic error. This situation is most likely to happen when the design, protocols, and equipment for the two experiments are the same. In that case, any systematic error in the two experiments is likely to be the same.

Ex. 1: Suppose your bathroom scale is out of calibration. Though you don't know it, it reads low by 2.5 lb. At the start of your diet, you get on the scale and it reads 146 lb. Three months later the scale still reads low by 2.5 lb. You get on it and it reads 137 lb. You compute the weight you lost, 146 lb - 137 lb, which is 9 lb. If the scale hadn't read 2.5 lb low, it would have read 148.5 lb at the start and 139.5 lb at the end, yet the difference would still be 148.5 lb - 139.5 lb, which is 9 lb.

The same can happen in research laboratories. For instance, wheat is grown in a plot treated with a fertilizer, and in a plot not treated. The wheat is harvested, dried, and weighed. Though the scale is out of calibration, making the measurements wrong, it doesn't matter in computing the differences in weights.

Ex. 2: Johannessen et al. (2005) analyzed a data set of radar altimeter height measurements made at fixed locations on the Greenland Ice Sheet for 11 consecutive years, 1992 to 2003. A height measurement is the height above sea level of the ice at a given location. The year-to-year change in height tells us, when it is positive, how much the ice cap grew; and when it is negative, how much it had shrunk.

It didn't matter that the measurements at each location contained a systematic error, shifting the measurements from the true heights. The researchers computed the difference in height measurements from year to year, and so the systematic error canceled out by subtraction. As for the actual numbers, the researchers computed the mean year-to-year change for the locations above 1500 meters, and for those below 1500 meters, and computed the standard errors. They reported that "an increase of 6.4 ± 0.2 centimeters per year (cm/year) is found in the vast interior areas above 1500 meters. . . . [and] below 1500 meters, the elevation-change rate is -2.0 ± 0.2 cm/year. . . ."

A kind of situation where systematic errors do not cancel out. When the raw measurements in the two experiments are shifted from their true values by systematic errors that differ in size, the difference in the raw measurements (or in their means) is affected by the systematic errors. This situation is most likely to happen when the design, protocols, or equipment for the two experiments are different. In that case, the size of the systematic error in each experiment is likely to be different.

Ex.: Suppose you read a report of an experiment where a researcher made measurements of some quantity, and read another report of another experiment where another researcher made measurements of the same quantity. The two experiments are not likely to have the exact same design and protocol, and hence the sizes of the systematic error in the two are likely different. The measurements made with one experiment will be systematically shifted more or less than the measurements made with the other experiment. In taking the difference in the means of the two sets of measurements, the effect of the systematic errors will not cancel out. In case the two sets of measurements are systematically shifted in the same direction, there will be some cancellation, but not perfect cancellation. The worst case is when the two sets of measurements are systematically shifted in different directions. Then instead of the systematic errors partially cancelling, they add.

Ex.: A biologist wants to know the difference between the size of an elk herd in a certain area this spring and the size last fall. The biologist does an experiment that measures the size this spring, and locates a report of an experiment that measured the size last fall. It happens, suppose, that the two experiments have different designs and protocols. It's likely then that the measurements are systematically in error by different amounts. Only in the case where both experiments have no systematic error will the difference between fall and spring measurements be the true difference.

Another kind of situation where systematic errors do not cancel out. The same systematic error is in both experiments, but the raw measurements in the experiments are used in nonlinear calculations to produce derived measurements. By this, the derived measurements in one of the experiments are disproportionately shifted relative to the derived measurements in the other experiment. Hence the difference in the derived measurements (or in their means) is affected by the systematic error.

Ex.: Following Tennekes' (1996) book, *The Simple Science of Flight*, suppose we make two sets of raw measurements on a herring gull (of course then releasing it to the wild). For the first set, we measure its weight, W, in newtons, and its wing surface area, S, in square meters. From W and S we calculate its minimum flying velocity, V, in meters per second, for staying aloft at sea level. This is the velocity necessary for the lift it develops to equal its weight, so that it flies level, neither rising nor dropping. Next we fatten it up, measure its weight, note that its wing surface area is the same, and compute V, its minimum flying velocity. We subtract the first velocity from the second, and the difference tells us how much faster the gull has to fly when it is heavier.

Tennekes gives the equation for a bird's minimum flying velocity as V = sqrt ((W/S) / 0.38), where "sqrt" indicates the square root. The value of 0.38 is a constant that is related to air density at sea level and to a wing angle of attack of 6 degrees, a representative value for long-distance flight.

Let's first take the case where the gull's weight is measured on a scale that contains no systematic error:

Before we fatten it up, we measure its weight, getting W = 9.4 newtons. And we measure its wing surface area, getting S = 0.181 square meters. With these raw measurements, we calculate the minimum flying velocity to stay aloft: V = sqrt ((W/S) / 0.38) = sqrt ((9.4/0.181) / 0.38) = 11.69 m/sec (26.15 mph).

After we fatten it up, we measure its weight, getting W = 10.81 newtons. And we measure its wing surface area getting, as before, S = 0.181 square meters. With these raw measurements, we calculate the minimum velocity to fly at to stay aloft: V = sqrt ((W/S) / 0.38) = sqrt ((10.81/0.181) / 0.38) = 12.54 m/sec (28.05 mph).

Neither experiment has systematic error in it. The difference in velocities is 12.54 m/sec - 11.69 m/sec = 0.85 m/sec (1.9 mph). When the gull is heavier, it has to fly faster, and this tells us how much faster.

Now comes the case where the before and after weights are measured on a scale that contains a systematic error. Specifically, the scale measures 1 newton lower than true weight. When we go on to calculate the difference in velocities, we will not get 0.85 m/sec, as we got above. This will demonstrate that the difference in velocity depends on the systematic error; i.e., the effects of the systematic error do not cancel. Here's the demonstration:

Before we fatten it up, we measure its weight, getting W = 8.4 newtons (1 newton less than it really weighs). And we measure its wing surface area, getting, as before, S = 0.181 square meters. With these raw measurements, we calculate the minimum velocity it needs to fly at to stay aloft: V = sqrt ((W/S) / 0.38) = sqrt ((8.4/0.181) / 0.38) = 11.05 m/sec (24.72 mph).

After we fatten it up, we measure its weight, getting W = 9.81 newtons (1 newton less than it really weighs). We measure its wing surface area, and it is the same as before, S = 0.181 square meters. With these raw measurements, we calculate the minimum velocity to fly at to stay aloft: V = sqrt ((9.81/0.181) / 0.38) = 11.94 m/sec (26.71 mph).

The difference in velocities is 11.94 m/sec - 11.05 m/sec = 0.89 m/sec (1.99 mph).

To summarize, when there is no systematic error in the before and after measurements, the difference in velocities is 0.85 m/sec. With the systematic error in the before and after measurements, the difference in velocities is 0.89 m/sec. If the agreement seems close, it is only because of this specific example. For other cases, with other raw measurements, and other equations containing stronger nonlinear effects, there could be substantial disagreement.

As the next example shows, it isn't true that systematic errors fail to cancel only when nonlinear derived measurements are used in the comparison. Nonlinear reactions of experimental material may cause systematic errors not to cancel.

Ex.: Suppose we want to investigate the effect of a novel fertilizer on the growth of rye. In the pre-measurement stage we prepare soil for growing potted plants of rye in a greenhouse. The treatment group of potted plants will receive the fertilizer; the control group won't. However, for both groups we make the same systematic error in preparing the soil. Specifically, we had only three-fourth the amount of soil moisture that protocol calls for. Suppose the plants in the two groups react differently to this, as they could. For the period of the experiment, the difference in the growth of biomass in the treatment and control groups will depend on the systematic error. Had it not been made, with plants in both groups having the correct soil moisture, the difference in growth of biomass would be another value.

To summarize this section: Systematic error will not cancel out where there is a nonlinear effect, and/or where the sizes of the systematic errors differ in the two experiments. What should you do when you foresee that the effects of a systematic error in a comparative experiment you are planning will not cancel? Try to plan the experiment in a way that minimizes systematic error, making it be nearly zero.

2.10 How to minimize systematic error

To minimize the systematic error in planning and carrying out an experiment, I recommend taking five steps:

Step 1: In planning an experiment, try your best to identify all sources of systematic error, and to come up with ways to minimize them. To be as thorough as possible, do this sequentially by the three stages discussed in section 2.4: the sampling stage, the pre-measurement stage, and the measurement and calculation stage. (If your experiment lacks a pre-measurement stage, so much the better for minimizing systematic error.)

Step 2: Seek advice from other researchers. Ask for help in identifying sources of systematic errors you may have missed. Ask for suggestions on how to minimize them. Revise your plan as need be. Pay your research colleagues and acquaintances to carefully and thoroughly go over your preliminary plan. People work most enthusiastically and attentively when they're getting money for it. Be generous. In the words of the old saying: If you pay peanuts you get monkeys. Two days' wages per person seems reasonable. Where does the money come from? Put it in your research proposals for grant funding, as a necessary item.

Step 3: Give a pre-project seminar to colleagues (and graduate students if you are at a university). Ask whether they notice any sources of systematic error that remain. Tell them you will give $5 for every one they find. If good suggestions result, incorporate them into your plan. (A colleague of mine took this approach for finding typographic errors in the draft of a textbook he wrote. He assigned the draft for class reading, offering a dollar for each typographic error to the first student who found it. He told me that when he made the assignment he'd have bet the draft was errorless, but the students proved him wrong.)

Step 4: If possible before doing the full experiment, do a pilot experiment. Discovering sources of systematic error by thinking in an armchair has its limits. You'll likely discover some sources you overlooked by planning and running a pilot experiment. You may think of additional sources of systematic error that neither you nor your colleagues had noticed. If you do, revise your plan as need be.

Step 5: When carrying out the final plan, pay close attention to doing exactly what it specifies to do. The enemy is mental lapses which let systematic error slip in; be on guard throughout.

Keep in mind that sources of systematic error can be staring you in the face while you are blind to them. Thomas Levenson put it as The Scarlet Pimpernel Principle: "One will always fail to notice the most obvious mistakes, for the mind denies the possibility of the truly egregious error that the eyes can clearly see." The Scarlet Pimpernel Principle is why it is advisable to have others go over your preliminary plan. What are inconspicuous systematic errors to you may be blatant to others. (Confirmation of The Scarlet Pimpernel Principle occurs when authors submit manuscripts to publishers, and the authors would bet the manuscripts are free of spelling and punctuation errors, yet always the editors uncover dozens. In one instance I know of, an author's book was published with his name misspelled on the cover because he failed to spot the error on the cover's proof.)

We now leave the subject of minimizing systematic error in experiments. We take up the subject of minimizing random error in making a statistical inference – or what is the same, the subject of determining the proper sample size.

2.11 How to determine the proper sample size for experiments

You are, let's assume, going to replicate an experiment n times to estimate a confidence interval or to test a statistical hypothesis. What value of n should you use? This is what we mean by the proper sample size.

For determining the proper sample size, collaborate with a statistician in planning your experiment. For, even an experienced mountaineer wouldn't dream of setting off to climb the

Matterhorn without an expert guide. With both collaborating, the chance of success is greatest. Just so, even a researcher experienced in statistical methods should not go it alone in research studies without a collaborating statistician.

Unless you are the exception, your knowledge of the methods of statistics is from one or two college courses that glossed over sample size determination. Have the statistician calculate the proper sample size. The statistician is a specialist in knowing the right equations to use and how to use them correctly. (Besides, collaborating with a statistician will brush up and increase your knowledge of statistics, paying dividends in your future research.)

How to determine the proper sample size for computing a confidence interval.

For the sake of discussion, let's assume you want to estimate a confidence interval for μ, the mean of a population. (The discussion will hold in principle for estimating a confidence interval for any parameter of a population, such as its standard deviation, σ.)

Along the lines suggested in the following examples, tell the collaborating statistician the values of two specifications you want your experiment to meet: (1) the percent confidence and (2) the allowable error (both defined below). Ask the statistician:

Ex.: "What is the proper sample size for my experiment so that I can get a 95-percent confidence interval on the true mean quadrat biomass of cheatgrass, μ, with an allowable error of 0.6 kg?"

Ex.: "I'd like a 95-percent confidence interval, with an allowable error of 8 $\mu g/m^3$, for the true mean amount of PM-2.5 in the air on January 4 at my study site. What's the proper sample size to take?"

Ex.: "How many adults do I need to randomly sample in Youngstown, Ohio, so that I can estimate a 95-percent confidence interval on the true mean credit card debt of Youngstown adults, with an allowable error of $200?"

Once the statistician knows this information – the percent confidence and the allowable error – and has estimated the variance of the measurements in the population you'll be randomly sampling, the statistician will compute the proper sample size for you.

Let's take a minute to refresh your understanding of percent confidence and allowable error. The key to understanding them is to think about the collection of all possible random samples from a population, rather than think about the specific random sample you take.

Percent confidence. Imagine you could repeatedly take random samples of, say, size $n = 100$ from a population, again and again, forever, and with the data of each sample compute a 95-percent confidence interval. Of these confidence intervals, the mean μ will be within 95 percent of them, and be outside 5 percent of them. In other words, 95 percent of them will be truthful, and 5 percent will be lies which "say" they contain the mean μ when they don't.

In life, however, when you take one random sample and compute one 95-percent confidence interval, there is a 95 percent chance that the one you computed is one of the truthful ones and μ really is within its interval. Conversely, there is a 5 percent chance that the one you computed is one of the untruthful ones, with μ really outside its interval.

Allowable error. As before, imagine you could repeatedly take random samples of, say, size n = 100 from a population, again and again, forever, and with the data of each compute a 95-percent confidence interval. Of these confidence intervals, their widths differ. Relatively few are very wide, relatively few are very narrow, with most in between. There is some average width of these confidence intervals.

In setting the allowable error, E, decide the widest average width of the possible confidence intervals that would suffice for the purpose of your research, divide it by two, and the result is the allowable error, E.

Equivalently, E can be defined as how close on average you would like the estimate, $\bar{X}$, to be to μ: $E = |\mu - \bar{X}|$. The two vertical lines indicate the absolute value of the difference of μ and $\bar{X}$ (i.e., if the difference is negative, make it positive.) Assuming you could randomly sample again and again, forever, with a given sample size n, E is the average closeness of $\bar{X}$ to μ.

The size of the allowable error you should specify depends on what you will use the confidence interval for. Is it to be used to conceive a research hypothesis? Is it to be used to test a research hypothesis? Is it to be used in a predictive computer model? Is it to be used in making a management decision? In addition to the particular use, it depends on your field. Medicine, for example, usually requires narrower confidence intervals than does educational research. At any rate, the smaller you set the allowable error, the narrower the average of all possible confidence intervals will be.

Why don't researchers specify an allowable error that is near zero? The proper sample size would doubtless be so large it would be unaffordable.

Think back now to asking the collaborating statistician, "What is the proper sample size so that I can get a 95-percent confidence interval on the true mean quadrat biomass of cheatgrass, μ, with an allowable error of 0.6 kg?" This means two things. (1) The researcher wants the imagined set of all possible confidence intervals to be such that 95 percent of them $((1 - \alpha) \times 100)$ will contain the true mean, μ. (2) The researcher wants the imagined set of all possible confidence intervals to be such that their average width is twice the allowable error, 2×0.6 kg = 1.2 kg. The researcher believes that 1.2 kg is the widest average width that will just suffice for the purpose of the research. Knowing that $\alpha = 0.05$ and that E = 0.6 kg, the statistician computes the proper sample size. Let's say the computation comes out to be $n = 32$.

Next, the researcher takes a random sample of 32 biomass measurements of cheatgrass, and from the measurements computes the confidence interval. This confidence interval is one from the set of all possible confidence intervals having the required 95-percent confidence and having the average allowable error E of 0.6 kg.

Comment: Imagine you are upstairs at night and you hear a noise downstairs. You listen for further noises. At some point you've heard enough further noises to call the police. Having any more data would not change your decision to call. By the same token, the proper sample size is sufficient for the purpose of your research. To choose tighter specifications, producing a larger sample size, would overly meet the purpose. Take the money saved from not overly meeting the purpose and spend it on other parts of your research, such as minimizing systematic error. Accordingly, the idea of a proper sample size goes against the idea that you can never have too much data, i. e., too large of a sample size.

Comment: If you had a dollar for every researcher who misunderstands what a confidence interval means, you'd be a millionaire. Let me demonstrate the typical misunderstanding with a passage I spotted in the October 16, 2006, issue of *Barron's*. There, a researcher at the U. S. Bureau of Labor is quoted as saying: "The 90-percent confidence interval on the monthly change [in total employment ranges] from -330,000 to 530,000 (100,000 +/- 430,000). The figures do not mean that the sample results are off by these magnitudes, but rather that there is a 90-percent chance that the 'true' over-the-month change lies within this interval." The researcher at the Bureau of Labor has it wrong. There is no 90-percent chance that the 'true' over-the-month change lies in the quoted interval of -330,000 to 530,000. Rather, the correct interpretation is that if all possible random samples of the given size were taken, each of which would give a 90-percent confidence interval, then 90-percent of the possible confidence intervals would contain the true over-the-month change. The quoted confidence interval is just one of the possible confidence intervals. There is a 90-percent chance that it is among the possible confidence intervals that contains the true monthly change, and a 10-percent chance that it is among the possible confidence intervals that don't contain the true monthly change.

How to determine the proper sample size for testing a statistical hypothesis.
Pretend you are going to test a statistical null hypothesis that the true mean, μ, is zero, against the alternative hypothesis that it is not zero. To know the proper sample size for making the test, tell the statistician the three specifications you want the test to meet: (1) the significance level, (2) the effect size, and (3) the power of the test. Use words along the lines of the following examples:

Ex.: "I'll be randomly sampling a population of teenagers. At the $\alpha = 0.05$ level of significance, I want to test the null hypothesis that the correlation coefficient between their I.Q. and the hours per week spent watching TV is zero in the population, this against the alternative hypothesis that the correlation coefficient in the population is at least 0.50. That is, the effect size is 0.50, as any smaller value will not be scientifically meaningful. Last, I want the power of the test to be 80 percent. What is the proper sample size for meeting these specifications?"

Ex.: "I'm going to randomly sample a population of people, and put them on a new kind of diet for five weeks. Right before the diet and right after it, I'll measure each person's weight. Then I'll compute the difference between each person's before-weight and after-weight. At the $\alpha = 0.05$ level of significance, I want to test the null hypothesis that the diet would have no effect if it were given to the entire population of people, against the alternative hypothesis that the diet would produce an average loss of weight of at least 5 pounds per person in the population. That is, the effect size is 5 pounds, as any smaller weight loss will not be scientifically meaningful. Last, I want the power of the test to be 80 percent. Please calculate for me the proper sample size for meeting these specifications."

Once the statistician knows the three specifications – the significance level, the effect size, and the power of the test – and has estimated the variance of the population(s) you'll be sampling, the statistician will be able to compute the proper sample size for you.

Now let's take the diet example and in detail discuss the three specifications. The symbol μ stands for the mean weight loss for the population. You will test the null hypothesis, H_o: $\mu = 0.0$ pounds, against the alternative hypothesis, H_A: $\mu \geq 5.0$ pounds. To do the test you will randomly select a sample of people, n in number, measure the difference in before and after weight for each, and compute the sample mean, $\overline{X}$, the average difference in before and after weight for the n people in the sample. The collaborating statistician will compute the proper value of n for you, after you tell her/him the values of the three specifications.

Significance level. In the event the null hypothesis, H_o, is really true, meaning the diet has no effect if given to all the people in the population, the significance level, α, is the probability of making a Type I error. This is the probability that, due to random error in the experiment, the test will lead you to reject the null hypothesis when it should not be rejected. Researchers usually set α at 0.05. Other values you will less often come across in published articles are $\alpha = 0.1$ and $\alpha = 0.01$.

Effect size. The effect size, ES, is the smallest difference in the variable you are measuring that would be a meaningful difference in terms of the research. In deciding that ES = 5 lbs, you have decided that an average difference of 5 or more pounds weight loss is a meaningful difference.

Power of the test. Statisticians define the symbol β to be the probability of making a Type II error in a statistical test. A Type II error is made whenever the alternative is actually true but the statistical test leads you to accept that the null hypothesis is true. For example, if the diet really would produce a mean weight loss of at least 5 pounds in the population, but on the basis of the test you conclude it would not, you would have made a Type II error. Now, the power of the test is defined as the complement of the probability of making a Type II error, with the power expressed as a percent. Hence the power of the test is $(1.0 - \beta) \times 100$. If you specify 80-percent power, then the percentage chance of making a Type II error is 20 percent (i.e., β is 0.2).

In the diet experiment, with the value of α set at 0.05, the effect size set at 5 pounds, and the power set at 80 percent ($\beta = 0.2$), the statistician will compute for you the proper sample size. Let's say it is $n = 42$ people. Here's what this means: Imagine a million separate diet experiments with a random sample of 42 people, and with each experiment you test the null hypothesis against the alternative hypothesis. And suppose the case is that the diet would really have no effect if given to the entire population; i.e., the null hypothesis is really true. Well, 5 percent of the time – approximately 50,000 of the 1,000,000 – you would make a Type I error, mistakenly concluding the diet would have an effect in the population.

Now, imagine another million experiments with the same specifications as before: α set at 0.05, the effect size set at 5 pounds, and the power set at 80 percent ($\beta = 0.2$), making the sample size be 42 people. But this time suppose the alternative hypothesis is true, i.e., the diet really has the stated effect of a 5 or more pounds average weight loss in the population. Well, 20 percent of the time – 200,000 of the 1,000,000 – the test will lead you to make a Type II error, mistakenly concluding that the diet would have no effect if given to the entire population.

What often happens is that with the specifications you set – the values you set for α, the effect size, and β – the computed sample size n that meets them is so large it is prohibitively costly to take. In such cases, the statistician will vary the specifications around the values you choose, computing a table of sample sizes for select values of α, of the effect size, and of β, and let you decide which revised specifications to use. From my experience, when cost considerations cause researchers to loosen the specifications they had hoped to meet, it is the power of the test they lower (i.e., they increase β), resulting in a smaller sample size. Of all experiments ever done, I would wager that far fewer meet, say, 80-percent power ($\beta = 0.2$) than, say, 50-percent power ($\beta = 0.5$). Cost considerations force researchers to accept a lower power than they'd like.

2.12 How to reward your collaborating statistician

How do you entice a statistician to want to collaborate with you in designing your experiments and determining the proper sample sizes? Reward the person well with money and recognition. In your research proposals for grant funding, include a statistician to be employed, and don't skimp on the pay.

Besides $$$s, people obviously work best when they are recognized for their work. In your reports and talks on your completed research, acknowledge your collaborating statistician's contribution. Rodbard (1982) explains: "In the absence of authorship or any other visible incentive, the statistician is most likely to provide minimal input; . . . Consulting statisticians often have little to show for their efforts, at least in terms of bibliography; . . . hence, the overall extent and quality of statistical consultation tend to suffer." Rodbard recommends that the statistician be made a coauthor, or at least given prominent mention with a by-line and phrases such as "with statistical analysis by X," or "with the statistical consultation of X." Why do this? One, it's a matter of fairness. Two, your research will benefit, as will science.

All professional associations that publish research journals ought to have at least one statistician on their manuscript review panel. One that does is the editorial office of the *Journal of the American Medicine Association*, which employs a statistician to examine submitted research manuscripts to determine whether or not the research has followed proper statistical protocols. Researchers, knowing that the statistical soundness of their experiments will be scrutinized by an independent statistician, with the possibility that publication will be denied after the research has been done, will be attracted to collaborate with statisticians in planning their experiments.

Res. Ex. 2.2: Estimating the average speed of vehicles

This is a sketch of a fictitious research example illustrating several topics covered so far in this chapter: the planning and execution of a three-stage experiment, with the findings given as a confidence interval and a statement of accuracy. Were the research real, the plan and execution of the experiment would have to be spelled out down to the last detail.

Here we go. Some teachers at Adams school believe that during school hours the mean speed of vehicles on the one-way road past the school exceeds the 25 mph limit. After their call to the police department, it assigns officer Alice to investigate. She plans an experiment to estimate μ, the mean speed of vehicles passing the school during school hours.

As part of her plan, Alice decides that I* (the largest inaccuracy that is acceptable) should be 3 mph. After she has run the experiment and made a statement of accuracy giving the value of i, if it happens that i is less than or equal to 3 mph she will consider the amount of systematic error in the experiment is acceptable. But if i exceeds 3 mph she will decide that the amount of systematic error in the experiment is unacceptable, and on this alone the experiment should be considered unreliable.

The sampling stage of Alice's plan:

(1) Collaborate with a statistician to determine the proper sample size. This is the number of vehicles (sample units) to be in the random sample. In determining the sample size, Alice sets the following specifications. She sets the allowable error, E, at E = 0.75 mph. This is how close the estimated mean speed of the sampled vehicles, $\overline{X}$, should on average be to the actual mean speed, μ, of all the vehicles passing the school on the day the sample is taken. And she sets the percent confidence at 95 percent. From this, let's suppose, she and her collaborating statistician calculate that the sample size is $n = 324$ vehicles to be sampled at random from the population.

(2) Prepare 1,600 numbers, numbered 1, 2, 3, . . . , 1,599, 1,600. The first car that passes on the day of sampling will be assigned the number 1, the second the number 2, and so forth. From the 1600 numbers select at random 400 of them, 76 more than $n = 324$. The numbers in this list indicate which cars to sample, measuring their speeds.

   Why 1600 numbers? From past data Alice knows that at most 1,600 vehicles could pass on any school day. And why select 400 of the 1,600, rather than 324, the proper sample size? This is to be safe; she doesn't want to get in a situation where she has sampled 324 vehicles before the end of the school day. That would under-represent late afternoon drivers, and they might drive slower or faster than morning drivers (which would be a source of systematic error). Thus the actual sample size she takes may be more than 324, i.e., up to at most 400. If it turns out to be more than 324, the estimating process will be more precise than it needs to be. That will be ok; costs aside, there's no harm in exceeding the specifications for precision.

(3) At random, pick a school day for doing the experiment. Alice, let's suppose, randomly picks April 9.

The pre-measurement stage of Alice's plan:

(1) Calibrate the radar gun so it will accurately measure the speeds of vehicles (an uncalibrated radar gun would be a source of systematic error).

(2) Use an unmarked car and wear street clothes (a police car and/or uniform might alert drivers to the experiment, prompting some to drive slowly, a source of systematic error).

(3) Park along the street (radar readings taken from a moving car can cause the car's speed to be added to or subtracted from the speeds of vehicles being clocked, a source of systematic error).

(4) Be sure the car's heater/air conditioner fan is off (a running fan motor can produce erroneous radar gun readings, a source of systematic error).

(5) Be well practiced in using the radar gun correctly (this helps avoid inaccurate readings such as "panning errors," a source of systematic error).

The measurement stage of Alice's plan:

(1) For each vehicle in the random sample, measure its speed with the radar gun.

(2) Have an assistant record the measurements.

This, then, is the plan for Alice's experiment. Next comes her execution of the plan:

On April 9, beginning at 8:00 a.m., she starts counting vehicles. Whenever the count matches a number on the list, she measures that vehicle's speed and her assistant records it. Suppose with about half an hour of the school day left, she has measured the speeds of 324 vehicles. So she continues using the list, and at the end of the school day has measured the speeds of 31 more vehicles. This gives her $n = 355$ measured speeds ($324 + 31 = 355$).

With the 355 measured speeds, she computes the average, and it is $\overline{X} = 27.9$ mph. Next, she computes the 95-percent confidence interval for the average speed per vehicle, μ, in the population of vehicles that passed the school. It is:

$$27.9 \pm 0.4 \text{ mph, or } 27.5 \leq \mu \leq 28.3 \text{ mph.} \qquad (2.7)$$

The width of the confidence interval reflects the amount of random error in her experiment.

Along with this, she estimates that the inaccuracy from systematic errors in the experiment is no more than $i = 1.5$ mph. Since this is less than $I^* = 3$ mph, she decides that the amount of systematic error in the experiment is tolerable.

In her report she writes the estimated vehicle speed of 27.9 mph is accurate within ± 1.5 mph. This means that although she was careful to minimize systematic error in her plan, and in the execution of it, a residual of systematic error was present, with a net effect of at most ± 1.5 mph.

And in her report she concludes that quite probably the average speed of all vehicles traveling past the school exceeded the 25 mph limit. Her reasoning is as follows. If there wasn't any systematic error in the experiment, i would be zero; the estimated speed, $\overline{X} = 27.9$ mph, would be perfectly accurate. Under this proviso, the confidence interval (eq. 2.7) justifies her conclusion, as it lies completely above the speed limit of 25 mph. Moreover, the worst case of systematic error is - 1.5 mph, which if removed from $\overline{X} = 27.9$ mph lowers it at most to a mean speed of 27.9 - 1.5 mph, or 26.4 mph. Because this is above the limit of 25 mph, her conclusion of speeding is still justified under the worst estimated case of systematic error.

Her findings reveal that quite probably the speed limit was exceeded that day. Further, she believes that day is typical of school days. She recommends that the city traffic department put up more prominent signs calling attention to the school zone and the speed limit, hoping this would solve the speeding, bringing it down to 25 mph or less.

2.13 Experiments that do not involve sample units

To this point in this chapter, the discussion about minimizing random error and systematic error has centered on experiments that involve sample units. Recall the role of sample units. There is a population of a finite number, N, of sample units, such as people, animals, plots of land, or days of the year. And there's a measurable property of interest on the sample units, such as income per person, weight per animal, number of saguaro cacti per plot, or PM-2.5 parts per million per day, and so forth. From the population of sample units we select n of them at random. Sometimes it is necessary to prepare the selected sample units for measuring the property of interest, and sometimes preparation is unnecessary. Either way, we make n measurements, i.e., replicate the experiment n times. The total random error in such experiments comes from random error due to sampling, from random error in preparing the samples (if they have to be prepared), and from random error in making the measurements. Similarly, the total systematic error is a combination of the individual systematic errors in each stage of the experiment.

Now, experiments that do not involve sample units have only a measurement stage. They are concerned with measuring fundamental properties. An example is of a physicist who makes n measurements of the light transmission efficiency in a fiber optics strand, and from the measurements estimates a confidence interval on the true transmission efficiency for the infinitely large population of possible measurements that could be made on the strand.

Another example is a triangulation experiment to estimate the distance from Earth to the star Zeta Ophiuchi. The experiment consists of measuring: (1) the angle from Earth to Zeta Ophiuchi when Earth is on one side of its orbit around the sun, and (2) measuring the angle from Earth to Zeta Ophiuchi six months or so later when Earth is on the opposite side of its orbit.

With the two angle measurements and a third measurement – the straight-line distance between the positions of Earth when the two angle measurements are made—the distance to Zeta Ophiuchi can be mathematically derived. The experiment involves no sample units. There is just Earth and Zeta Ophiuchi, and the geometry with respect to them and more distance objects. Nazé (2006) mentions that astronomers have replicated the triangulation experiment, estimating the mean distance between Earth and Zeta Ophiuchi to be 463 ± 46 light-years (the $\pm$ amount is the standard error of the mean).

Whether or not an experiment involves sample units, all of the recommendations in this chapter apply, such as using standard constructs, minimizing systematic error, and minimizing random error.

Experiments in fields of the physical sciences tend not to involve sample units. Just look at research articles in journals dealing with planetary science, physics, astronomy, and chemistry. And experiments in fields of the life sciences and social sciences tend to involve sample units. Just look at research articles in fields such as food science, forestry, soil science, wildlife science, biology, botany, cell biology, ecology, toxicology, atmospheric sciences, environmental sciences, geology, geophysics, meteorology, oceanography, paleontology, epidemiology, immunology, veterinary science, and all the fields of engineering.

2.14. Three questions that researchers describing their experiments must answer

Researchers, when they describe their experiments in research manuscripts and seminars, are responsible for answering three questions without being asked. As for reviewers of manuscripts, it is their responsibility to make sure the researchers have answered the three questions, under penalty of denying publication. As for members of audiences listening to researchers describe their experiments, it is their responsibility to make sure that the researchers can satisfactorily answer the three questions.

Here are the three in the form of asking a researcher who has presented a seminar but failed to address them, and it is now the question-and-answer period: (1) "How did you properly determine the sample size(s) you used?" (2) "How did you minimize the systematic error in your experiment?" (3) "What is the size of the minimized systematic error in your experiment (i.e., the statement of accuracy), and why is it tolerable?"

2.15 About induction

In talking about inferential statistics in this chapter, we have been talking about induction. The "duct" in induction runs from the facts of a sample to the larger population of facts from which the sample facts came. You know induction: "This fish lives in water, that fish lives in water, everywhere I look the fish are living in water. I'm going to generalize and believe that all fish – the whole population of fish – live in water." This is induction.

We use induction casually every day. I'm thinking of when we take a bottle of milk that's been in the refrigerator for two weeks, and want to decide whether it is spoiled or not. We pour a glass and take a tiny sip. It tastes ok. Just to make sure, we turn the rim, take another sip, and it tastes ok. On a sample of two sips we conclude that all the milk in the glass is ok.

We share our capacity for inductive thinking with all of the mammals. Bertrand Russell (1974) pointed out that induction goes on instinctively in the primitive faculties. A cat will inductively sample its safety on its new owner's lap. Sample 1: "I was there for a second, jumped right off, and was safe." Sample 2: "I was there a bit longer and was safe again." [. . .] Sample 9: "I was there for an hour, safe all the time." At some point the cat automatically makes the generalization: "I'm always safe there."

Hustlers in games such as shooting pool and playing golf rely on their opponents' tendency to make inductive generalizations. Gambling in the early games for small stakes, a hustler will let the opponent win. At some point the opponent decides, "I'm a lot better than this person is." Then the hustler suggests the stakes be raised. About the worst kind of wickedness is using induction to gain a person's or an animal's trust, then hurting it.

Whether in the everyday sense or in the formalized sense of the methods of inferential statistics, induction is a method of generating knowledge. In the coming chapters of this book are other methods, built on "duct," for generating knowledge: retroduction (chapter 3), hypothetico-deduction (chapter 4), and deduction (chapters 7 and 8).

2.16 About qualitative facts

To this point in this chapter we have discussed quantitative facts. Quantitative facts are measurements of degree. They involve numbers representing magnitudes. Now let's discuss qualitative facts. Qualitative facts are measurements of kind. They involve categories. In research, qualitative facts may be as important as quantitative facts. In certain circumstances, for instance, the qualitative fact that it snowed may be more important to a meteorological researcher than the quantitative fact of the depth of the snow.

Phenomena. Phenomena are qualitative facts. Archeologists excavate the remains of a Neanderthal and observe the phenomenon of flower petals buried with the body. Research psychiatrists view Van Gogh's mental breakdown as a phenomenon compared to the phenomenon of a normal mind.

Aurora borealis is a phenomenon. The construct named "aurora borealis" can be connotatively defined as "colored and white flashing lights in the night atmosphere, particularly north of the Arctic Circle." Accordingly, astrophysicists can observe the Arctic sky on a December night and measure whether the aurora borealis – the phenomenon – is present or absent.

The qualitative classes of phenomena may be more than two, as with cloud types, coded as, say, 5 = cirrus, 4 = cirrocumulus, 3 = cirrostratus, 2 = nimbostratus, 1 = cumulus, 0 = cumulonimbus. A climatologist who records the sky's condition as "4" has made a qualitative measurement of the phenomenon "cloud type."

Categorical properties of quantitative measurements. It commonly happens that a researcher does an experiment that produces raw quantitative measurements, processes the measurements to produce a derived quantitative fact, and then from it extracts a property of interest, a qualitative fact. Some examples:

An aquatic ecologist wants to know which streams in a set are similar with respect to pollutants in their waters. The experiment involves measuring the amounts of various polluting chemicals in the streams. The ecologist puts the quantitative measurements into software for cluster analysis, and it produces a qualitative map of classes of similarly polluted streams.

An economist measures the annual income of several thousand households in an area and summarizes the measurements with a frequency distribution. From the frequency distribution, which is quantitative, the economist extracts a qualitative fact of interest: that the distribution is bimodal.

At times in all fields, researchers' interest lies mainly in the qualitative shape of frequency distributions. If the distribution is symmetric, that may mean one thing; if skewed to the right, another thing; if skewed to the left, another. Or if the shape of the curve formed by plotting one variable against another is S-shaped, that may mean something that wouldn't have been meant by a curve that isn't S-shaped.

Of course, where qualitative facts are extracted from quantitative facts, researchers aim to minimize systematic error and random error in the underlying experiments. For if the experiments producing the quantitative facts are too much distorted by errors, the qualitative facts extracted may be wrong.

To sum up, much research involves qualitative facts, of which there are two kinds. One is phenomena, such as cloud types. One is categorical properties of quantitative measurements, such as shapes of statistical distributions, shapes of mathematical functions fitted with statistical regression techniques, and maps of classes produced with multivariate techniques, e.g., cluster analysis or factor analysis.

2.17 The place of experimental facts in P-plane diagrams

Fig. 2.1 is an illustrative P-plane diagram. P-plane diagrams are ways of picturing knowledge and the methods of discovering it. **P-plane** is short for "plane of perception." Below the P-plane is nature, i.e., what is "out there" beyond our minds. Above the P-plane is the **C-field**, short for "the field of constructs." Constructs, recall from section 2.8, are well-defined constructed concepts that can be measured, giving facts, quantitative or qualitative. Scientists define constructs to connect the mind, represented by the C-field, with what is "out there" in nature.

The particular P-plane diagram in Fig. 2.1 is adapted from one appearing in Margenau (1961). Constructs are shown as double lines, capped with a circle, connected with the P-plane. They are conduits from nature into our minds, and are defined by the measuring instruments and protocols for taking measurements. Experiments produce facts that appear at the top of constructs, in the circle part. From there they are logically processed in the mind, using theories and knowledge, producing deduced facts in other parts of the C-field. The circles in the C-field that are not connected with double lines are constructs called **isolates**, because they are isolated from direct connection with nature. They are well-defined concepts but cannot be directly measured.

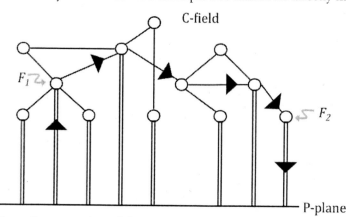

Figure 2.1. A P-plane diagram, adapted from Margenau (1961), illustrating a circuit of empirical verification through the C-field.

To closely paraphrase Margenau: A circuit of empirical verification (following the arrows in Fig. 2.1) starts at fact F_1, which undergoes a transition through the C-field. F_1 might be the observation of the position of a comet at a certain time. That part of the C-field through which the circuit moves might be Newton's law of gravitation, and the final transition to the construct at the far right of Fig. 2.1 might be the predicted position of that same comet at some future time. If the predicted position proves to tolerably agree with the actual position, F_2, that is a successful passage through the circuit. When a set of constructs and isolates is successfully traversed by a great number of such circuits, the set is said to be verified, and the constructs and isolates are said to correspond to reality.

What if a predicted fact fails to tolerably match the corresponding measured fact? It means something in the circuit traversed is wrong. If that happens, it is a stimulus to scientists to find out what it is and try to fix it, making the circuit work.

In chapter 3 we will show how scientists conceive research hypotheses, representing them as isolates in the C-field. In chapter 4 we will show how scientists test research hypotheses, and how P-plane diagrams help in understanding the tests. In chapter 5 we will show how scientists establish cause and effect, from one construct to another in the C-field. And in chapters 7 and 8 we will show how, from existing knowledge in the C-field, scientists deduce new knowledge.

3

How to conceive and circumstantially support research hypotheses

No lights came on at sunset. . . . The machinery had failed again, or the power had been deliberately turned off, or the power station had been captured by the rebels. **V. S. Naipaul,** *A Bend in the River*

Well put, Mr. Naipaul! The passage illustrates the "fact method" for conceiving hypotheses. It starts with an interesting fact: No lights came on at sunset. It ends with conceiving a hypothesis that possibly explains the interesting fact. In Mr. Naipaul's passage we see the "fact method" applied three times, giving three hypotheses, each capable of explaining the interesting fact. Hypothesis 1: The machinery had failed. Hypothesis 2: The power had been turned off. Hypothesis 3: The power station had been captured.

The "fact method" is one of three methods for conceiving research hypotheses. The other two are the "question method" and the "subject method." I invented their names, guided by wanting names that are easy to keep in mind. In brief, the three are as follows:

With the **fact method**, you note an interesting fact, or several related interesting facts. The fact(s) may be qualitative, such as a phenomenon, or may be quantitative, such as a sample mean. Then you think up a possible answer – a research hypothesis – that, were it true, would explain the interesting fact.

Galileo Galilei, in a notable example, used the fact method. Through his telescope he saw the interesting fact of four dots of light lined up around Jupiter, nightly changing their positions. To explain the fact, he hypothesized they were moons of Jupiter.

With the **question method**, you come upon an interesting question about a condition or a process of nature. You think the question up, or another researcher does. Either way, it stimulates you to work out a possible answer – a research hypothesis.

Friedrich August Kekulé, in a notable example, used the question method. As the story goes, he had been pondering the question, "What is the shape of a molecule of benzine?" Stimulated by this, one night he dreamed of a snake taking its tail into its mouth. This suggested the answer. He hypothesized that the shape is a ring.

With the **subject method**, you think about a subject and at some point the thinking suggests a possible truth – a research hypothesis – about a condition or a process of nature.

Leo Szilard, in a notable example, used the subject method. For some time, he had been thinking about the physics of neutrons hitting atoms of uranium-235. One morning, walking to work, the mechanistic hypothesis of a chain reaction suddenly dawned on him, where neutrons split atoms of uranium-235, releasing neutrons that split even more atoms of uranium-235, releasing even more energy and neutrons, and so on.

(Scientific discovery often starts with having a research hypothesis, and usually we can't conceive one without applying one of the above three methods. I say "usually" because there is one exception. It is worth mentioning but not discussing further in this book. It is accidental discovery, without a hypothesis. Though there are a few cases – Antoine Henri Becquerel's accidental discovery of radioactivity is one – Weinberg (2003) notes that "the proportion of scientific discoveries achieved by accident is not as large as many people think.")

About the three methods it's natural to say: "There's nothing to them. What practical value are they to me?" Well, suppose you want to conceive a hypothesis. What should you do? Stimulate your mind, and for this there are three approaches. (1) Find an interesting fact(s) and start from that. (2) Ask an interesting question and start from that. (3) Think long about a subject and start from that. It's important to have the three ways of starting, three stimulants, and three ways of igniting an Aha! By no means are the three substitutes for one another. Some research hypotheses are easier to conceive by starting from an interesting fact, some by starting with an interesting question, and some by regularly immersing yourself in thinking about a subject.

The fact method is the bread-and-butter method of research. Therefore, this chapter gives it the most space. Still, learn and practice all three methods. Researcher A who is well practiced in the three is more likely to contribute more to science than Researcher B who is well practiced in just two of the methods. Researcher B is more likely to contribute more to science than Researcher C who is well practiced in just one of the methods. And Researcher D, not well practiced in any of the methods, is bound to contribute the least of all.

Because the fact method is so widely used, the philosopher of science, N. R. Hanson, has studied its use in research. His name for the fact method is the **method of retroduction**. It's the name we will favor in this book, because it is well established in the literature of the philosophy of science. It is also called **the retroductive method**. Still, at times we will refer to it as "the fact method."

I've organized the chapter into three groups of sections, as follows:

1 – UNDERSTANDING THE SIMPLE FUNDAMENTALS OF THE THREE METHODS

Section 3.1 explains what research hypotheses are and how they differ from statistical hypotheses. Section 3.2 explains the retroductive method (the fact method) in the way we all use it to conceive hypotheses about everyday affairs. Section 3.3 explains how to use the retroductive method to conceive research hypotheses. Section 3.4 explains how to use the question method and the subject method to conceive research hypotheses.

2 – COVERING TOPICS THAT APPLY TO ALL THREE METHODS

Section 3.5 explains how researchers use the methods to recursively conceive and evaluate research hypotheses. Section 3.6 is about gathering such conclusive circumstantial evidence for a hypothesis that it amounts to a "smoking gun," indicating the hypothesis is almost certainly true and all competitors are almost certainly false. Section 3.7 explains mechanistic research hypotheses. Section 3.8 explains the original method of multiple working hypotheses, and the modified method. Section 3.9 pictures, with a P-plane diagram, the role of research hypotheses in science.

3 – ROUNDING OUT YOUR VIEW WITH RESEARCH EXAMPLES

Section 3.10 contains a list of diverse examples that will round out your view of conceiving research hypotheses.

3.1 What research hypotheses are

To understand research hypotheses it helps to compare them with statistical hypotheses. Their fundamental difference is this: Statistical hypotheses are about facts, measurements. Research hypotheses are about how nature is or how nature works.

Here is a statistical hypothesis: The difference in the mean tail length, μ_1, of all the house cats living in the U.S., and the mean tail length, μ_2, of all house cats living in Italy, is zero. This is called a null hypothesis, "null" indicating the value of zero. In symbols, the null hypothesis about tail length is: H_0: $\mu_1 - \mu_2 = 0.0$. A commonly stated alternative hypothesis is that the difference between μ_1 and μ_2 is not zero. In symbols, this alternative hypothesis about tail length is: H_A: $\mu_1 - \mu_2 \neq 0.0$.

In contrast, here is a research hypothesis: "Cats evolved tails because tails have survival advantage; specifically, when cats fall from heights, tails help them to right themselves and land on their feet." Notice that this, unlike a statistical hypothesis, is not about measurements. Rather, it explains how nature works.

You can measure cats' tails until you are blue in the face, and it will get you no closer to understanding why they have tails. To understand why requires imagining a reason for their having tails, and the imagined reason is a research hypothesis.

Classifying research hypotheses. The four main classifications of research hypotheses are independent of one another. They are the following. (1) Some research hypotheses provide explanations, some provide answers. (2) Some research hypotheses are qualitative, some are quantitative. (3) Some research hypotheses are simple, some are mechanistic. (4) Some research hypotheses can be conceived relatively quickly, some take months or longer. Consider each:

CLASSIFICATION BY EXPLAINING OR ANSWERING

Research hypotheses that explain: Curious about an interesting fact, we retroductively conceive a research hypothesis to explain it. We may want to explain *why* a certain fact is as it is. Ex.: From the fact that the risk of having dyslexia is greatest for children born in May, June, or July, a medical researcher goes on to conceive several hypotheses that explain why. Or we may want to explain *how* something works. Ex.: From the fact that Monarch butterflies migrate from the U.S. to breeding areas in Mexico, an ecologist goes on to conceive explanations of how the Monarchs reach these areas without getting lost.

Research hypotheses that answer questions: Such research hypotheses provide answers to actual or implied questions of the likes of *Where?*, *When?*, *Who?*, *What?*, and *Is?*. To illustrate:

From the fact that turquoise beads, dug out of a prehistoric living site in Arizona, called Snaketown, have a characteristic chemical composition, an archaeologist proposes a research hypothesis that specifies where the source mine for the turquoise was located.

From the fact of the lines of a handwritten poem discovered on a piece of paper laid in a book that appears unopened for a hundred years, a linguist proposes a research hypothesis that specifies when it was written. Or the linguist proposes a research hypothesis that specifies who wrote it.

From the fact of an unusually shaped bronze tool uncovered in an archeological dig, a historian of technology conceives a hypothesis answering what the tool was used for.

From the facts of the orbital properties of a planet in another solar system, an astrophysicist conceives a research hypothesis stating that there is water on the planet. Used in this way, the verb "is" answers questions of existence. Another: a biologist proposes that living microorganisms exist on Mars.

Besides positing the existence of something (e.g., water on a planet), a research hypothesis may posit the existence of a cause-effect relation between one thing and another. Some illustrations: "Free-drifting icebergs cause hot spots of chemical and biological enrichment in the southern ocean." "Cancer stem cells drive tumor growth." "Genetic diversity in termite colonies enhances colony productivity and long-term fitness."

CLASSIFICATION BY QUALITATIVE OR QUANTITATIVE

Another way to classify research hypotheses is by whether they are qualitative (stated in words) or quantitative (stated in mathematics). Look for research hypotheses in reports of research in, say, the journal *Science* or the journal *Nature*. You'll find research hypotheses that are expressed in words, like the one above for why cats evolved to have tails. And you'll find research hypotheses that are expressed in mathematics, like an astrophysical research hypothesis in the form of a mathematical model of how a star explodes into a supernova.

CLASSIFICATION BY SIMPLE OR MECHANISTIC

Simple research hypotheses have few parts. The qualitative ones can be written in a dozen or so words, such as "Acute intermittent porphyria was behind Vincent van Gogh's mental breakdowns." The quantitative ones can be concisely expressed, such as "$E = mc^2$ "(it was a research hypothesis for a while after Einstein conceived it, then became regarded as confirmed knowledge).

Mechanistic research hypotheses are complicated with numerous related parts. They require many words or equations to state. Here's the sort of thing: A team of researchers conceives a research hypothesis explaining how northern Africa, once lush with vegetation, became the Sahara desert. The research hypothesis is a mechanistic model, expressed in words and flow diagrams, linking ideas of global terrestrial ecosystem theory with ideas of climate theory.

CLASSIFICATION BY HOW LONG IT TAKES TO CONCEIVE THE RESEARCH HYPOTHESIS.

Some research hypotheses can be conceived in minutes, hours or at most a week, and some take much longer. As a case in the first class, it didn't take researchers a great deal of time to conceive several possible explanations for Vincent van Gogh's mental breakdowns. One research hypothesis posits bipolar disorder. Another posits poisoning from drinking absinthe. As a case in the second class, it took James Watson and Francis Crick more than a year to conceive the research hypothesis that a gene is a sequence of DNA arranged in a double helix.

There's a lesson in this. If you have unsuccessfully tried for a week or so to conceive a research hypothesis, that doesn't mean it's impossible and you should give up. A place to see this is the research of Loftus and Arnold (1991). Going beyond the above two "van Gogh hypotheses," it took them some time to arrive at the research hypothesis that van Gogh suffered from acute intermittent porphyria, an inherited metabolic disorder.

3.2 The retroductive method as practiced in everyday affairs

You already know the retroductive method. It's ingrained in us. I want to emphasize this with several everyday examples. This will make it clear that its use in science is natural.

To begin with, recall the method. A fact (or set of facts) captures your interest. This stimulates you to figure out a likely explanation for the fact, i.e., a hypothesis that, if it was true, would account for the fact.

Illustration 1. After sleeping soundly through the night, you wake to the fact that the street is wet. Reasoning retroductively, you conceive a possible explanation. Hypothesis 1: It rained during the night. For if it had rained, the fact of the wet street would be explained. Naturally you can't definitely conclude that it did rain; there are other ways the street could have gotten wet.

Reasoning retroductively again, you conceive a different possible explanation for the fact. Hypothesis 2: Sometime during the night the water main broke. For if it had broken, the fact of the wet street would be explained. Again, you can't definitely conclude that it did break, only that if it broke that would explain the street's being wet.

You keep on reasoning retroductively, getting more hypotheses such as runoff water from fighting a house fire up the street, and an airliner jettisoning waste water. You end with a set of hypotheses, possible explanations.

The words retroductive and retroduction are not listed in common dictionaries. Words special to the philosophy of science never are. "Retro" means "against or opposite to." Retroduction is a form of inference that goes in the direction opposite to reasoning with deduction. With deduction, we can reason that rain during the night will definitely make your street wet. With retroduction, we can reason that your street being wet means possibly it rained during the night.

Illustration 2. Jane says to Sue, "Barbara called and said Jim's just been diagnosed with colon cancer." The fact: Jim has colon cancer. Jane retroduces a hypothesis that explains the fact. She tells Sue, "It's all that meat he ate." Sue, answering, has retroduced her own hypothesis, "It's his genes. Barbara told me once that Jim's dad and aunt died of colon cancer, and they were mostly veggie eaters." Later, Jane and Sue decide that the two hypotheses are best combined into a third about diet and genes.

Retroduction is the main way detectives reason. Clues are found, suspects are hypothesized. In the Sherlock Holmes stories, Arthur Conan Doyle often has it wrong in writing that Holmes made a deduction. Actually, Holmes mostly made retroductions. Consider the difference. This is deductive reasoning: "Watson, we'll leave this murder-crazed man, Mr. Morgan, in the house tonight with Sarah. I deduce that tomorrow morning we will see the fact that Sarah is dead." In contrast, this is retroductive reasoning: "Watson, we found Sarah dead this morning. From this fact I retroduce that Mr. Morgan possibly was in the house with her last night and killed her."

3.3 How to use the retroductive method to conceive research hypotheses

Slightly paraphrasing Hanson (1958), he lists the steps of the retroductive method as:
1. Some interesting fact (or set of facts) F is observed.
2. F would be explained as a matter of course if the research hypothesis, H, were true.
3. Hence there is reason to think that H might be true.

Interesting facts. An interesting fact excites you and others with an itching desire to know what accounts for it. If you can conceive a hypothesis that explains the interesting fact, then you have an interesting hypothesis. If you can further support the interesting hypothesis with circumstantial evidence, and/or corroborate it with a hypothetico-deductive test (covered in chapter 4), then you have discovered interesting knowledge. Often, interesting knowledge turns out to be important, a shoo-in for being publishable in a top research journal. Prospects are good for researchers building on important knowledge, and citing it in their research articles. Prospects are good for researchers saying after important knowledge is published, "It's a great paper. This is going to have a strong impact on the way we think about. . . . This is going to open the way for advances in our field."

Uninteresting facts don't arouse curiosity. To take one example, it's an uninteresting fact that in the average old house there is more dirt per square foot in the attic than in the kitchen. A scientist who succeeds in retroducing a reliable hypothesis explaining this uninteresting, and hence unimportant, fact is sure to get from journals one rejection slip after another.

To sum up, what's interesting in science usually proves to be important to science; what's uninteresting usually proves to be trivial. It's wise to be always on the lookout in your field for interesting facts that have yet to be explained. Spot them in journal articles, in talks with colleagues, and in your experimental research. Try to conceive research hypotheses explaining them. Don't waste time trying to conceive hypotheses that explain uninteresting facts.

Some research examples of using the retroductive method. Each of the following research examples focuses on an interesting fact, and one or more research hypotheses are retroduced, each capable of explaining the interesting fact. Keep in mind that where there is more than one hypothesis in an example, they may have come from separate researchers working independently. And keep in mind that the examples present only the use of retroduction, and not subsequent work by researchers to evaluate the credibility of the hypotheses by using circumstantial evidence and/or the hypothetico-deductive method.

For brevity, where it cannot lead to misunderstanding we will sometimes write "hypothesis" in place of "research hypothesis."

Res. Ex. 3.1: Why are our two lungs asymmetric?

From his dissecting of human corpses, Leonardo da Vinci surely knew that our two lungs are not mirror images. The fact is, the right lung has three lobes, and the left lung has two. As to why, Parker (1997) appealed to the necessity for efficient packing of organs in the chest cavity. Hypothesis: The left lung has one fewer lobe to make room for the heart.

Res. Ex. 3.2: Why did Mozart curse?

The fact is that Mozart's letters to his cousin Maria Anna Thekla Mozart are spotted with obscenities. Simkin (1992) recognized a possible reason. Hypothesis: Mozart suffered from Tourette's syndrome, a disease in which sudden uncontrollable outbursts of cursing are often a symptom.

Res. Ex. 3.3: Why do some landslides "have legs?"

Once they hit level ground, landslides tend to run unbelievably long distances. Petry (1992) and Woodard (1992) independently realized that this fact would be explained if the following hypothesis was true. Hypothesis: An avalanche pounding down a steep slope sets the relatively level ground ahead vibrating. Upon reaching the vibrating ground, the avalanching rock-fragments are bounced along, reducing friction. At the same time, Richards (1992) realized that the fact would be explained if a different hypothesis was true. Hypothesis: Like truck brakes that fail from overheated brake drums, the sliding rocks in an avalanche heat up, reducing the friction between them and the ground, letting them flow more easily. Naturally both mechanisms – heated rocks, and moving along vibrating ground – acting in concert can be made the basis of a third hypothesis.

Res. Ex. 3.4: Why do many species of snakes eat their shed skins?

Weldon et al. (1993) suggest two plausible explanations for the fact of snakes' eating their shed skins. One is for a physical benefit. Hypothesis 1: The skin is a source of protein. One is for a behavioral benefit. Hypothesis 2: Eating the evidence of their presence protects snakes from predators stalking them. The two hypotheses could be combined to make a third about the joint benefits, protein and protection.

Res. Ex. 3.5: Might genes play a role in divorce?

McGue and Lykken (1992) noted the fact that the divorce rate for married couples is highest when both partners have divorced parents, is next highest when one of the partners has divorced parents, and is lowest when neither partner has divorced parents – even though the environments and living situations of the couples differ widely. To McGue and Lykken, this fact admits this likely explanation: Hypothesis: There is a genetic influence on divorce risk, apart from any environmental influence.

Res. Ex. 3.6: Why do women menstruate?

Disease is at the center of Profet's (1993) hypothesis to account for the fact that women menstruate. Hypothesis: Menstruation evolved in women because it cleanses the uterus of sperm-borne bacteria, reducing the chance of disease. Strassmann (1996) instead put energy at the center of her hypothesis to account for the fact. Hypothesis: In the main, menstruation evolved because it saves energy. The monthly growth and retreat of endometrial tissue in the uterus takes less energy than keeping the blood-filled endometrium ready to receive an embryo.

(Facts and questions concerning biological function are regularly addressed from the perspective of evolution. Witness The European Society for Evolutionary Biology, whose stated objectives are to "support the study of organic evolution and the integration of those scientific fields that are concerned with evolution: molecular and microbial evolution, behaviour, genetics, ecology, life histories, development, paleontology, systematics and morphology.")

The retroductive method is easily abused by applying it once, considering the retroduced hypothesis as the best explanation of the fact, and making no effort to search for circumstantial evidence against it and for it. Such is a recipe for getting it badly wrong. A make-believe research example emphasizes the point:

Res. Ex. 3.7: Could an economy run on pay toilets?

That it could is the scatological hypothesis retroduced from an interesting fact in Leo Szilard's science fiction story, "Report on 'Grand Central Terminal'" (Szilard 1961). The story tells of aliens who land on Earth, find the human race has been destroyed, and go into Grand Central Terminal and retroduce a hypothesis about the nature of human civilization. Quoting from the story, the aliens find an interesting fact: Small cubicles which must have "served as temporary shelter for earth-dwellers while they were depositing their excrements, with the door of each locked by a rather complicated gadget" containing "a number of round metal disks, each of which had written on it the word Liberty." From this the aliens hypothesize: "By barring earth-dwellers from depositing their excrements unless they sacrificed a disk on each occasion, they were made eager to acquire such disks, and that the desire to acquire such disks made it possible for them to collaborate in co-operative efforts which were necessary for the functioning of their society."

3.4 How to use the question method and the subject method to conceive research hypotheses

Recall the two methods:

The question method: You entertain an interesting question about a condition or a process of nature. Possibly you thought it up, or possibly another researcher did. Either way, the question stimulates you to work out a possible answer, a research hypothesis. For instance, suppose a soil scientist becomes fascinated with the interesting question, "Can competition among soil bacteria be overcome?" The scientist goes to work trying to conceive and support a research hypothesis that, if true, would answer the question. Perhaps it is this: Hypothesis: The strength of soil pH affects competition, and at some high value of pH competition among soil bacteria will stop.

The subject method: You are thinking about a subject and at some point the thinking suggests a possible truth – a research hypothesis – about a condition or a process of nature. For instance, suppose an oceanographer is thinking about the microscopic animals living in coral reefs, and about their role in promoting the health of reefs. This leads to the thought of bacteria infecting the microscopic animals, killing them. Then a possible truth comes to mind, namely: Hypothesis: The wetsuits of divers, when not disinfected after dives, can harbor bacteria, fatally infecting microscopic animals.

In one sense, the question method and the subject method verge on being the same. With the question method, you are conscious of the question that stimulates you to answer it with a research hypothesis. Yet conceivably with the subject method, you are unaware that in your unconscious mind you have asked a question and worked out an answer, a research hypothesis. Then at some moment the research hypothesis appears consciously as though it didn't come from a question. For instance, asking yourself unconsciously the question, "Are many public saunas a source of disease?," might prompt you to become aware of the following hypothesis: "The aerosol in public saunas is conducive to transmitting tuberculosis."

In two other senses, the question method and the subject method differ. One sense is that certain hypotheses are more easily discovered with one of the methods than with the other, although which method that is varies among cases. Take for illustration the hypothesis that increased CO_2 levels in the atmosphere caused glaciers of the last Ice Age to retreat simultaneously, causing their relatively quick end. My hunch is that this hypothesis would be more likely to arise from a researcher thinking on the subject of global warming than from a researcher entertaining the question "Was the end of the last Ice Age relatively quick or drawn out?"

The other sense where the two methods differ lies in the different ways we force them to work. To force the question method to work, be a regular questioner and in time you will hit upon good research questions, the starting point to working out tentative answers, hypotheses. To force the subject method to work, regularly immerse yourself in thinking about the knowledge in your field, and in time new research hypotheses will come to mind.

Illustrations of the question method. Question: Where is dark matter in the universe located? Hypothesis: Dark matter resides inside stars.

Question: Does intelligent life exist beyond our solar system? Hypothesis: It is almost certain that there is intelligent life elsewhere in the universe.

Question: Were cereals and legumes the earliest domesticated crops in the Near East? Hypothesis: Fig trees were cultivated before cereals and legumes were cultivated.

Question: How widespread is the practice of voodoo in Louisiana today? Hypothesis: Ten to fifteen percent of the people living in New Orleans are likely to have practiced voodoo. Statewide, five percent of Louisianians have probably practiced voodoo.

Question: What is the moon's crust made of? Hypothesis: The lower crust is richer in iron, titanium, and thorium than the upper crust.

That these examples involve simple hypotheses is incidental. Mechanistic hypotheses might in certain cases be conceived as well, e.g., a mechanistic explanation of how dark matter could reside inside stars.

The question method may lead to either specific questions or general questions. Specific questions arise at the specific level of science, while doing specific research projects, and so they are numerous. The question "How widespread is the practice of voodoo in Louisiana today?" is specific. In contrast, the question "Is ours the only universe?" is general. Many general questions in science have already been thought of. They are, so to speak, on the books.

General questions act to get the research ball rolling in a general area of inquiry. They beget specific questions, which leads to hypothesizing specific answers, some of which pass testing and become specific knowledge. So it is that the accumulated specific knowledge amounts to general knowledge. In Francois Jacob's (1982) words: "In fact, the beginning of modern science can be dated from the time when such general questions as 'How was the universe created? What is matter made of? What is the essence of life?' were replaced by more modest questions like 'How does a stone fall? How does water flow in a tube? What is the course of blood in the body?' Curiously enough, this substitution had a quite unexpected result. While asking general questions led to very limited answers, asking limited questions turned out to provide more and more general answers. This is still valid for present-day science."

Illustrations of the subject method. All of the hypothesis illustrated above that originated with the question method could originate with the subject method as well. I mean, you are reading and thinking, not conscious of asking any question, and into your mind pops this hypothesis: Dark matter resides inside stars. Or this hypothesis: It is almost certain that there is intelligent life elsewhere in the universe. Or this hypothesis: Fig trees were cultivated before cereals and legumes were cultivated.

3.5 How to recursively conceive and evaluate research hypotheses

There are three steps to the recursion for conceiving and evaluating research hypotheses.

Step 1: Conceive a plausible hypothesis.

Step 2: Rigorously search for all negative and positive circumstantial evidence that bears on the hypothesis' probable truth.

Step 3: If step 2 fails to yield a well-supported hypothesis, repeat steps 1 and 2. Do this until you have hit on a well-supported hypothesis, one with no evidence against it and much for it.

Circumstantial evidence includes circumstantial facts and/or circumstantial knowledge. Among circumstantial knowledge are laws, and generally accepted theories and analogies bearing on the hypothesis' probable truth.

We can't combine steps 1 and 2 into just one step, conceiving a well-supported hypothesis at one go. For plausible hypotheses are conceived intuitively, without critical scrutiny. We have therefore to rigorously see how they stand up to circumstantial evidence. It tends to happen that on close examination, faults are found in them, and it's back to the drawing board, starting at step 1 again.

At step 1, commonly the retroductive method (the fact method) is used. It is well suited to being applied repeatedly to come up with different hypotheses that explain a given interesting fact. However, it could happen that in a series of cycling through steps 1 and 2, the question method or the subject method might be the basis of one of the cycles.

To illustrate the recursive process, pretend we are listening in on an ecologist's thoughts. "Now there's a weird fact. Milkweeds secrete a lot more nectar at night than in the day. Not too smart, when there are ten times more pollinating insects active in the daytime than nighttime. What company runs its TV ads while viewers are asleep? But, hey, wait a minute. Bakers are up at three in the morning baking bread and rolls, ready and fresh for customers first thing in the morning. Why not milkweeds too? It makes sense." At this point the ecologist has retroduced the following possibly true hypothesis: Milkweeds secrete most of their nectar at night so that an accumulation is ready in the morning to attract butterflies and bees. The ecologist (at step 2) looks for circumstantial evidence, and finds none that goes against the hypothesis. And later, circumstantial knowledge supporting the hypothesis flashes into the ecologist's mind. "Great. The hypothesis is consistent with the theory of biological economics. The energy cost to the milkweed plant is lowest, in terms of water loss, if it secretes nectar at night when it isn't needed. Neat. Sensible. I apologize, milkweeds."

Then a letdown: negative circumstantial evidence, initially missed, comes to light, with "Darn it. I think I remember so-and-so's paper arguing it's unlikely that the reason nectar accumulates at night is to be ready for the morning's butterflies and bees." So-and-so's paper is found. Forget the role of butterflies and bees; it's on to trying to retroduce another hypothesis that explains why milkweeds accelerate their nectar secreting at night. As it will turn out, moths browsing at night will have a central role in the next hypothesis (for details, see Morse (1985)).

Comment: As this illustration suggests, part of the support for research hypotheses may come from analogies with aspects of human living, as with the ideas about TV advertising hours and about bakers' hours.

Comment: Too bad that scientific reports almost never capture the exciting part of science: the recursive journey: the twists and turns of thinking, the ups and down of emotion. More young people would want to become scientists if more of them sensed the adventures of the chase that go on in scientists' minds and hearts.

The recursive process. Let me drill home the extreme lengths scientists will go with the recursive process that ends with step 3. I am thinking of astrophysicist Richard A. Muller, a former student of physicist Luis Alvarez. Muller (1985) says this: "Alvarez had taught me that most new ideas fail. One simply had to keep trying. One out of 10 ideas might be worth actually trying, and out of these, one out of 10 might lead to an important discovery. You need to have 100 ideas to have a chance at a real discovery. The important thing, the tough thing, is to keep on trying, to keep on having new ideas."

Muller practiced this advice to explain certain geological features in relation to gaps in the fossil record. He created the Nemesis hypothesis – the mechanistic idea that the sun has an undetected companion star, dubbed Nemesis, that every 26 to 30 million years nears the sun. As a result, its gravity pulls comets loose from the outer solar system, some spill into the inner solar system, and some of these hit Earth, kicking up debris, blocking sunlight, creating a risk to complex life forms. In arriving at the Nemesis hypothesis, Muller went through developing more than a dozen hypotheses and seeing how they fit with circumstantial evidence, until he hit on the Nemesis hypothesis that explained the extinctions of dinosaurs and fit all of the then available circumstantial facts and knowledge.

In evaluating hypotheses, "Pick the fruit from the lower limbs first" is a good strategy for assembling circumstantial evidence. It might be that certain circumstantial facts or knowledge are relatively easier to obtain, and your hunch is that they might allow you to quickly dismiss the hypothesis as false and get on with the business of finding one that will stand. When you think about it, detectives confronted with a list of suspects follow this strategy. Hoping to quickly pare the list, first they try to find out which suspects had no opportunity to commit the crime. For the remaining suspects who had opportunities, the detectives do what takes longer than examining opportunity; they assess each suspect's potential motives. Usually, following this order is more efficient than doing the reverse of assessing motives first, opportunities second.

An informal form of hypothetico-deduction goes on in the recursive process. With the strict form of the hypothetico-deductive method, we test a hypothesis by predicting (i.e. deducing) a fact that would exist if the hypothesis is true. Then we do an experiment and obtain the actual fact. Then we judge whether the prediction is true or false. Whereas within the recursive process of conceiving and evaluating hypotheses, an informal form of the hypothetico-deductive method goes on. We determine whether or not the hypothesis has consequences that agree with or are in conflict with existing circumstantial facts and knowledge. An instance in geology would be when a plausible hypothesis is conceived (at step 1) and it is noticed (at step 2) that a consequence of it contradicts knowledge of plate tectonics. Something has to go, and it must be the hypothesis. But if the consequence is consistent with knowledge of plate tectonics, this would lend some support to the hypothesis.

Circumstantial evidence and the strict form of hypothetico-deduction make a one-two punch. Besides collecting circumstantial evidence and using it to decide how credible a research hypothesis is, if possible strictly test the hypothesis with the hypothetico-deductive method. It may happen that a hypothesis with circumstantial support is further corroborated with the hypothetico-deductive method, in which case the hypothesis is very probably true.

More information on recursion. Much of a book of mine (Romesburg 2001) presents the recursive process by which scientists, artists, mathematicians, entrepreneurs, and indeed all creative workers create recursively, feeling their way along toward finding truth, gathering information, evaluating it, and changing course as required. The book presents samples of this from a variety of fields. Several are from science, including how Watson and Crick used recursion to envision the structure of genes.

As to the roles of intuition at step 1 and intelligence at step 2, the book discusses numerous creators, including Ingmar Bergman, who said: "I make all my decisions on intuition. But then, I must know why I made that decision. I throw a spear into the darkness. That is intuition. Then I must send an army into the darkness to find the spear. That is intellect."

Patience and thoroughness of thinking. Creating and supporting research hypotheses isn't done in a jiffy. I stress this to students when I teach my course in Best Research Practices. I give them an idea of the patience and thoroughness of thinking required by showing them research examples that border on the incredible. One concerns James Clerk Maxwell who, after studying the results of electrical experiments of Michael Faraday and of other experimenters, devised a research hypothesis, known today as Maxwell's equations. They are four, short, interrelated equations – you could write them on a postcard – that perfectly predict the behavior of electric and magnetic fields. What's more, they fix the idea that nothing can move faster than light. And I show my students a research article by Burbidge et al. (1957) in which the authors draw upon a mass of facts about quantum physics and nuclear cross-sections to conceive and support a research hypothesis that explains how stars make elements above hydrogen. I let my students weigh the article's 104 pages in their hands, and look through its bibliography of more than 200 references. Finally, I have them leaf through Prusiner's (1998) article, containing 347 references. It's an abbreviated version of his 1997 Nobel Lecture. In the article he explains the mechanistic research hypothesis he retroductively worked out for how abnormal protein molecules, called prions, cause transmissible spongiform encephalopathies, such as mad cow disease. Then I revive the students from their faint by explaining that most research hypotheses in articles from any current week's issues of *Science* and *Nature* are conceived with less extreme thought-work. Still, they as researchers should expect to work about at least two hundred hours to conceive and support a research hypothesis.

Some research examples. It's helpful to look at specific cases of researchers conceiving and supporting research hypotheses. Though surely the researchers went through many recursions, I will report only the hypotheses they finally landed on.

Res. Ex. 3.8: What causes dyslexia?

Dredging data, looking for ties between dyslexia and circumstances of one's birth, Livingston et al. (1993) came upon an interesting fact, a statistically significant association that the risk of having dyslexia is greatest for children born in May, June, or July. They proposed a reason and gave it support. Hypothesis: Viral diseases in pregnant mothers are responsible for dyslexia. Support: Dyslexic children developed in the womb during the period from December to April, when the exposure of their mothers to viral diseases, as influenza and measles, is greatest, which may affect the development of some fetuses' brains.

Res. Ex. 3.9: Why are Little Red Riding Hood's cape and hood red?

From Gardner (2000) and Hunter (2001) comes the following: In the 17th century, Charles Perrault wrote a story based on an old folk tale, calling it "Le Petit Chaperon Rouge" ("Little Red Cape"). Translated into English, the story is "Little Red Riding Hood." There's an interesting fact about it: Red, the color of her cape and hood, is in the story's title. Why not instead call the story "Petit Chaperon" in French, and "Little Riding Hood" in English? Why feature red?

Hypothesis 1: Red is featured for its sexual symbolism.

Supporting circumstantial evidence: The story has Freudian undertones. Red stands for a young girl's coming of age, her menstrual blood. And it stands for the hot-blooded wolf that gets in bed with her grandmother.

Hypothesis 2: Red is featured because it was the color people commonly wore.

Supporting circumstantial evidence: During the 17th and 18th centuries, red was a popular color for winter clothing. A warm color, it boosts the wearer's spirit while traveling about on grey, depressing, winter days. A safety color, red stands out against snow, increasing the chance of being rescued if lost. A cheap color, fabric needs only one pass through a red dyeing vat; most other colors need several.

Res. Ex. 3.10: What chronic illness was Tiny Tim suffering from?

Charles Dickens didn't say in "A Christmas Carol," but he may have seen a young person with the illness and adopted it for Tiny Tim. Lewis (1992) has gleaned a variety of facts relevant to Tiny Tim's condition:

Facts from the story: Tiny Tim's muscles would sometimes go weak. He wore leg braces and supported himself with a crutch. At times walking was too much for him, and his father carried him on his shoulders. He was pale, his teeth bad. He was unusually short, and yet had normal body proportions. His siblings were apparently healthy. Scrooge paid for his medical treatment after learning from the Ghost of Christmas Future that Tiny Tim would probably die before the next Christmas.

Facts about the environment: In the story's setting of London in the early 1840s, toxins such as lead and cadmium polluted the environment.

Facts from medical practice: Doctors at the time were in the habit of routinely prescribing fish oil, rich with vitamin D, and prescribing bicarbonate of soda.

To account for the facts, Lewis recursively retroduced a hypothesis, viz., Tiny Tim had renal tubular acidosis, a kidney disorder.

Lewis provided circumstantial evidence lending support to the hypothesis: Tiny Tim probably didn't have rickets because his brothers and sisters were apparently healthy. Although his symptoms were consistent with those of tuberculous osteomyelitis of the hip, knee, or spine, as well as those of hormone deficiencies, nothing could be done for these conditions at the time; Tiny Tim's getting well rules them out. Furthermore, occasional muscle weakness and weak bones are symptoms of renal tubular acidosis, and, although physicians at the time could not have known this, prescribing tonics of fish oil and bicarbonate compounds comes close to how the disease is treated today.

The next two research examples are from astronomy. Practically all of astronomy's explanatory knowledge comes through retroducing hypotheses and discovering those that circumstantial evidence and knowledge well support.

Res. Ex. 3.11: Why is the night sky dark?

On a clear night when the moon isn't visible, how can it be that the sky is dark? After all, the universe is thick with stars shining in all directions. People have long wondered about this, and in time it got a name, Olbers' paradox, after Heinrich Wilhelm Olbers, who wrote a paper on it in 1823. Various solutions have since been proposed. Croswell (2001) chronicled the following ones:

Hypothesis 1 (credited to Johannes Kepler): The universe is finite, and so it is wrong that there are infinitely many stars. Stars run out at some point, with a wall of darkness beyond. Looking out into space, we see mostly the wall of darkness.

Hypothesis 2 (credited to Jean-Philippe Loys de Chéseaux): Space is infinite, and although every line of sight from Earth outward hits a star, space is not perfectly transparent. Dust blocks the view to infinity.

Hypothesis 3 (credited to Edgar Allan Poe): The light from the really faraway stars has not yet had time to reach us. With time, the night sky will brighten.

Hypothesis 4 (credited to Edward Harrison): All of the stars in the universe together do not have enough energy to light the night sky. In support of this, Harrison calculated that for the night sky to be lit, every star would have to shine ten trillion times more brightly than it does. Or, put another way, there would have to be ten trillion times as many stars shining as brightly as the average star does. Today, astronomers regard Harrison's hypothesis as the primary reason for the dark night sky. That there is dust in space, and that light from really distant stars has not reached us, are lesser contributing factors.

Res. Ex. 3.12: Where is the nursery for comets?

The orbits of most comets are highly elongated about the sun (unlike the orbits of the planets, which are somewhat circular). That fact caught the attention of astronomer Jan Oort, and around 1950 he retroduced an explanation:

Hypothesis: Comets originate in a cloud of chunks of ice and rock located far out beyond Pluto (later known as the Oort cloud), 20,000 to 50,000 times more distant from the sun than is Earth.

Oort reached this hypothesis by calculating that only it could explain the highly elongated cometary orbits of several dozen comets he examined.

More recently, reports Kerr (2000), astronomers proposed a hypothesis related to Oort's:

Hypothesis: The Oort cloud is not spherical but is flat like a pancake, and comets with highly elongated orbits, such as Halley's, come from the part of the Oort cloud nearest Earth.

Circumstantial evidence in the form of computer simulations lent the hypothesis strong support. The astronomers tracked, for a simulated billion years, the motions of 27,700 computer-simulated comets that were influenced by the sun's gravity, by the four largest planets, and by passing stars. With the Oort cloud spherical in their simulations, highly elongated orbits were not produced. Only with the Oort cloud flat, and with the simulated comets' source in the part of the cloud nearest to Earth, did highly elongated orbits emerge.

Res. Ex. 3.13: How did we feed ourselves two million years ago?

Zimmer (2004) describes an application of the retroductive method by Dennis Bramble, biomechanics expert, and Daniel Lieberman, physical anthropologist. Two factual differences between humans and apes caught their attention.

Fact 1 concerns endurance running. The human species excels at running long distances. Seven hours and 51 minutes is the current record for running across the Grand Canyon and back (41.2 miles). Obviously the apes (gorillas, chimpanzees, orangutans. . . .) can't come close to that.

Fact 2 concerns bodily form. We have, but the apes lack, certain anatomic traits that make endurance running possible. We have short toes, stabilized arches, large heel bones, muscular rear ends, and industrial strength knees and ankle joints. We have long Achilles tendons rebounding us from stride to stride, wasting little energy. And we have a whiplash resister: the nuchal ligament, stretching from the base of the skull to the base of the neck. "Every time your heel hits the ground, your head wants to topple forward" (Lieberman). "It's an elastic band that has repeatedly evolved in animals that run. Apes don't have it" (Bramble).

To explain these facts, Bramble and Lieberman retroduced the scavenger hypothesis.

Hypothesis: Human evolved ability for long-distance running because that gave them an edge at scavenging.

That is, our early ancestors lived by scavenging for dead animals; they ran to places where they saw vultures gathering in the sky. By beating lions, hyenas, and vultures to a carcass, notes Lieberman, "you would have a lot of protein and fat at your disposal." And by beating other humans who were less fit for running long distances, the anatomic traits for endurance running emerged.

Expertise and experience. The other night I mentioned to my favorite layperson, my wife, the facts of the special anatomy of humans that led to the scavenger hypothesis. I asked whether she could think of any other hypothesis that could account for the facts. After a while she said, "The people were probably nomads. They walked long distances. Children have short legs, and it was run or be left behind. Children that had the beginnings of traits for endurance running were able to keep up, and that favored their survival and passing on the traits." Is this a sound competitor to the scavenger hypothesis? There's no way for her or me to tell; we're novices in anthropology. For one thing, she guessed the people were nomads. And she guessed the adults were heartless and wouldn't wait for the children. It would take an expert to bring rigor to bear, and know whether or not she is on to something.

There's a lesson here. Compared with scientists, novices find it easier to create explanations, for novices are prone to gloss over details. But in science, the devil is invariably in the details. Scientists have to know their business inside and out, part by part. That means knowing which details are significant and which are not. This is why no layperson ever (to my knowledge) won a prestigious scientific prize.

3.6 Assembling circumstantial evidence that amounts to a smoking gun

Perhaps you have (as I have) heard the belief that the only really credible research hypotheses are those that are corroborated with the hypothetico-deductive (H-D) method. The belief's wrong. Below, we are about to see examples of research where, although the hypothetico-deductive method was not used, you and I would bet a small fortune that the hypotheses are correct. They rest solidly on circumstantial evidence that amounts to a smoking gun. (Before the age of smokeless gunpowder, authors of murder mysteries could write such sentences as, "Your Honor, the victim was shot dead and not three feet away the suspect was holding a smoking gun.")

Real life provides situations where a single piece of circumstantial evidence forms a smoking gun. In 1921, in Hay-on-Why, England, there was a case of a solicitor who "would have got away with poisoning his wife had he been able to explain why the arsenic found in his possession had been divided into 20 small packets" (Maddox 2000).

This notion of a smoking gun carries over to research. Sometimes researchers are able to assemble circumstantial evidence that so strongly favors a certain hypothesis being true that there's no doubt about it.

Smoking-gun circumstantial evidence is most convincing when the number of possible hypotheses is finite and we are quite sure we are considering them all. Sayers (1941) explains that it is this way in detective stories, where a finite number of suspects are identified, one of whom did it. As the detective assembles more and more circumstantial evidence, certain suspects are ruled out, and certain others are more incriminated. The smoking gun appears when all but one are ruled out, leaving the one that must have done it. The same can happen in research, provided that the conceivable hypotheses are finite and few in number. Such is often the case with hypotheses about *where*, *when*, *who*, and *is*, as discussed in section 3.1. The following research examples demonstrate smoking-gun circumstantial evidence.

Res. Ex. 3.14: Who did it?

On June 27, 1930, in Amarillo, Texas, Mrs. A. D. Payne was driving the family car when an explosive planted in it went off, killing her. Mr. Payne, a prominent lawyer, usually drove the car the three miles to his office, only on this day he told his wife he was going to walk for the exercise, leaving the car for her use. After the funeral, suspicions in town settled on him but the local sheriff and his deputies, the police force, insurance company detectives, and newspaper reporters had no proof. To explain the fact that Mrs. Payne's car blew up killing her, the townspeople and authorities retroduced this hypothesis: Mr. Payne rigged the car with the explosive. Trouble was, the supporting evidence was too little to establish the hypothesis' truth to the degree necessary for getting a conviction in court.

Gene Howe, an Amarillo newspaper's editor, offered a $500 reward to anyone who could solve it. A. B. Macdonald, a reporter for the Kansas City Star, trained to ferret out news details, responded to the offer. His newspaper story (Macdonald 1930) of how he cracked the case won the 1931 Pulitzer Prize for reporting. Quoting from his story:

"I [Macdonald] reached Amarillo late the afternoon of August 2. Howe met me at the train. My first question was: 'What do you think of it?' 'I don't know,' he replied. 'Sometimes I think Payne killed her and again I think he did not. I am completely baffled. So are all the authorities. Ninety percent of the people here suspect Payne killed her, but there isn't a shred of proof.'" Macdonald went to work to find supporting reasons of circumstance and motive that would make the hypothesis very convincing. He discovered that Mr. Payne, despite appearing to love his family, and taking walks with his wife holding her hand in public, had secretly traveled out of town with his secretary. He discovered that Mr. Payne had cautioned his son, who survived the explosion, not to tell the authorities what had happened that morning. Macdonald got the boy to open up and say that after his mother had driven six blocks, the car filled with smoke. The boy said he told her, "Mother, that smells exactly like the smoke of that powder fuse I found in the backyard and burned." She answered, "Your father told me to drive faster if the car smoked."

Each supporting reason removed a degree of uncertainty from the hypothesis, until they together amounted to a smoking gun that made belief in the hypothesis unshakeable. Faced with this, Mr. Payne broke down and admitted he had done it.

Nobody wins a Pulitzer Prize for stating a hypothesis. Practically the whole town had done that. Pulitzer's are awarded for superior investigative reporting, for excelling at buttressing a hypothesis beyond the point of reasonable doubt. None in the town but A. B. Macdonald did that.

Res. Ex. 3.15: Where and when did Ansel Adams click the shutter?

About 30 years after Ansel Adams took the photograph, *Moon and Half Dome*, someone asked him what the date was when he took it, and where he was standing. He hadn't kept records and remembered only that it was in the late 1950s or early 1960s. This piqued Olson et al. (1994) to go sleuthing. With smoking gun exactness, they conceived a hypothesis answering when and where.

To pinpoint when, they gathered facts from the photograph, including the phase and altitude of the moon, and certain features on the moon's edge. On just a handful of dates bracketing 1960 could the moon have been positioned to account for its photographed phase and altitude. Further, they were able to eliminate some of the dates because particular features on the moon's edge, visible in the photograph, could not have been visible on the dates because of known lunar librations. To try to shorten the list of remaining possible dates, they examined weather records to learn the days when Yosemite Valley was not clouded over. Only one day was clear, December 28, 1960.

From what spot did Adams take the photograph? To pinpoint this, the researchers again gathered facts from the photograph: the geometry of certain terrestrial features in relation to the moon's position. Only one camera location gave the proper alignment of the terrestrial features with the moon. It was at the east end of Ahwahnee Meadow, 250 feet southeast of the stone gate on the road entering Ahwahnee Hotel. And as viewed from this position, only at 4:14 p.m. was the moon in relation to the sun and Half Dome as in the photograph.

Taken together, the facts they collected were so restrictive of possibilities that they produced a smoking gun, a soundly supported hypothesis, entirely believable. It's that Ansel Adams snapped the shutter on December 28, 1960, 4:14 p.m., plus or minus one minute, from 250 feet southeast of the stone gate on the road entering Ahwahnee Hotel.

Res. Ex. 3.16: Where did the 5,200 year old Iceman live?

In 1991 in the Alps, a 5,200 year-old mummified body of a man was discovered sticking out of a melting snowbank. An important question on scientists' minds was, "Where had he lived?" To answer it, Müller et al. (2003) measured three kinds of chemical isotopes in the Iceman's body, marking him with an isotopic signature. One kind of isotope was fixed in him when he was a youngster. A second kind was fixed in him when he was a teenager. The third kind was fixed in him when he was an adult.

The researchers found a good match of his isotopic signature with the isotopic signature provided by a certain place near the Alps. They found poor matches of his signature with the signatures of all other candidate places. This was a smoking gun. The question was answered with this hypothesis: The Iceman spent all his childhood and adult life in the Eisack Valley in north Italy, less than forty miles from the discovery site. As this verges on incontestable truth, we shouldn't call it a hypothesis, but rather consider it confirmed knowledge.

Res. Ex. 3.17: Who wrote the *Federalist Papers*?

Over a period of months in 1787-88 there appeared 77 unsigned newspaper essays that aimed to persuade citizens of New York State to ratify the U.S. Constitution. The 77 became known as the *Federalist Papers*; they and 8 related essays were published in a book, making 85. By the mid-1950s, most historians were certain of the identities of the authors of 73 of the 85: Alexander Hamilton authored 51, James Madison authored 14, John Jay authored 5, and Hamilton and Madison collaborated to write 3. There was no consensus of who authored the remaining 12, known as the disputed papers.

Later, Mosteller and Wallace (1964) and Bosch and Smith (1998), researchers independently following essentially the same approach, came to the conclusion that James Madison wrote all of the disputed papers. Their method: they collected data on the characteristic word frequencies and writing styles of Hamilton, Jay, and Madison, and on the word frequencies and writing style in each of the disputed papers. Madison's characteristic style well matched that of all the disputed papers, and Hamilton's and Jay's did not. A smoking gun, it was proof enough to hypothesize and confirm in one stroke that Madison was the disputed papers' sole author.

Research for identifying undisclosed authors of written texts is called stylometry. It is similar in principle to the research on the Iceman. That employed the matching of isotopic signatures. The *Federalist Papers* research employed the matching of word frequency and writing style signatures.

Another place to see stylometric research is the research of Jose Binongo, a stylometrician who identified the actual author of *The Royal Book of Oz*, the 15th book in Frank L. Baum's series of children's Oz books. The publisher printed Baum's name on the cover, but because the book appeared two years after his death, scholars had speculated that it was written by Ruth Plumly Thompson, a successor to Baum. To settle the question, Binongo collected samples of the writing of the Oz books known to be by Baum, and samples of the later Oz books known to be by Thompson. Binongo found that samples from *The Royal Book of Oz* were stylistically similar to Thompson's style, and dissimilar to Baum's. Klarreich (2003) describes Binongo's approach.

Res. Ex. 3.18: Do atoms exist?

The Greek philosophers were fond of applying the question method to come up with hypotheses about nature. To cite one case, to the question "Do atoms exist?" Democritus hypothesized that atoms do exist. His supporting reason was his belief that matter can be subdivided only so far; at bottom, there are atoms (atomos is the Greek word for indivisible). Aristotle didn't believe it. He hypothesized that matter can be divided infinitely, with no final smallest division, no atoms. For about 2,000 years, each hypothesis had supporters. Why did it take so long to resolve the problem? The microscope had to be invented.

One day in 1827, the Scottish botanist Robert Brown, looking into his microscope at pollen grains suspended in water, saw an interesting fact: particles in the solution, far tinier than the pollen grains, were jittering. Brown retroduced a hypothesis that the particles were something alive. This he soon decided was wrong. He (applying the hypothetico-deductive method) suspended mineral powders in water; they jittered too.

The jittering was ceaseless. Anytime people looked, day or night, January or July, it was there. It became known as Brownian motion, and it proved to be a smoking gun that meant the hypothesis that atoms exist is true. For, not long after 1900, Jean Baptiste Perrin and Albert Einstein independently retroduced what is surely the only possible explanation for the jittering: water molecules (H_2O packages of atoms) randomly banging into the tiny particles in suspension.

That's that. Speculation is ended. There is no other reasonable cause. Perrin qualitatively argued that if atoms "had no existence it is not apparent how there would be [Brownian motion]." To be doubly sure, Einstein created a quantitative statistical model of the random collision process of water molecules on tiny particles, and the model predicted detailed statistical properties of the jittering particles. In turn, Perrin experimentally determined the actual statistical properties, and they agreed with the predicted. (Chapter 4 presents the hypothetico-deductive method. But note this now: Together, Einstein's model and its prediction, and Perrin's experiment that determined the prediction is true, are a case of applying the hypothetico-deductive method.)

For more, see Lacina's (1999) quite readable article titled "Atom – from hypothesis to certainty." It tells of events from Democritus and Aristotle at the start, to Perrin and Einstein at the finish. I recommend the article for what it shows about the twists and turns of scientific inquiry, and how persistence wins the day.

By the way, ever wonder why Nobel Prizes in physics have been awarded to inventors of instruments? The tunneling electron microscope comes to mind. New instruments bring scientists new kinds of facts.

Res. Ex. 3.19: What does DNA look like?

That's the question that stimulated James Watson and Francis Crick to hypothesize the structure of DNA. They began by trying to build physical models of DNA. They cut out cardboard shapes of the nucleotides known to be in DNA, and played at arranging and rearranging them, looking for ways the shapes could be fitted together. By this they discovered the unique pairings of nucleotides that could be stacked in a way that agreed with relevant rules of chemistry and known properties of DNA. From there they went on to construct a model made of balls connected by rods, supported on jigs, stacking the paired nucleotides and companion molecules in a helical line. This fit the extant X-ray crystallographic data of DNA. Then came the day when unexpectedly staring them in the face was a likely truth they had not consciously designed into their model – a replicating mechanism for DNA to convey genetic information – a dead giveaway that the model was right. Their research provides a good illustration of (1) the question method of conceiving hypotheses, and (2) a smoking gun to pin down the truth of the hypothesis with certainty.

3.7 Mechanistic research hypotheses

Recall from section 3.1 that research hypotheses can be classified as simple or mechanistic. Simple research hypotheses have few parts and can be stated in a dozen or so words. Mechanistic research hypotheses have many parts and can't be stated in as few as a dozen or so words. To understand mechanistic research hypotheses better, let's look at two examples of them.

Res. Ex. 3.20: What causes Saturn's periodic white spot?

Since the late 1700s, astronomers are aware that every 30 years a large white spot forms on Saturn's surface. It also happens that it takes Saturn about 30 years to complete a revolution around the sun. This association between Saturn's 30-year revolution and the white spot appearing every 30 years is an interesting fact. To answer what might be responsible for it, Sanchez-Lavega et al. (1991) retroductively devised the following mechanistic hypothesis.

Hypothesis: Saturn's axis is tilted to its plane of revolution around the sun, giving it seasons. During one of these seasons the sun's rays fall almost perpendicularly on the region where the spot regularly appears. This causes a thermodynamic reaction in Saturn's gaseous depths: a growing vertical temperature difference in Saturn's troposphere, promoting a vertical rise of thermals from deep within, boiling over at the surface, creating the observed spot.

This, then, is a qualitative mechanistic hypothesis that is able to explain Saturn's white spot. With equal correctness we could call it a model of Saturn's white spot, understanding that "model" means "hypothesized model" and not "corroborated model."

Mechanistic hypotheses and their circumstantial support. With simple research hypotheses, the circumstantial evidence and reasons supporting them is noted outside of the hypotheses. With mechanistic hypotheses, it is usually different. The hypothesis is formed in a way that incorporates circumstantial evidence and reason favoring it, giving the hypothesis a degree of built-in support. Consequently, once we've conceived a mechanistic hypothesis, we shouldn't be upset that there is little or no circumstantial evidence and reason to be noted for it. Our next example is a good illustration of this.

Res. Ex. 3.21: How do we owe our existence to worms and snails?

The facts of interest are these: Existing worldwide are glacial-deposited rocks, laid down during a period that lasted from approximately 750 to 580 million years ago. They are covered with carbonate rocks, formed by precipitation from warm seawater. The first of these carbonate rocks began forming about 580 million years ago.

To account for these facts, Hoffman et al. (1998) conceived a qualitative mechanistic hypothesis called "the snowball Earth" hypothesis. According to it, about 770 million years ago, the tropical supercontinent Rodinia, containing all of Earth's land, broke into sections which are approximately the continents of today, and the sections began moving apart into the surrounding ocean. The breakup brought land areas, once far inside Rodinia's interior, nearer to ocean water. In turn, this brought increased rainfall on these land areas. The rainfall washed carbon dioxide out of the atmosphere. That decreased the greenhouse effect, which lowered Earth's temperature, which promoted the formation of large ice masses in the polar oceans.

The advance of this ice toward tropical latitudes fed on itself. The more ice there was, the more white color there was to reflect the sun's rays back into space, and the cooler Earth got. As ice formed in the lower latitudes, the sun's rays struck it more toward the perpendicular, increasing the efficiency of reflecting the rays into space. Besides, the thickening and advance of ice was helped along by the sun being dimmer then, giving off 6 percent less heat than today.

When the temperature spiral-down came to rest, Earth's land was covered with ice, and its seas frozen over to depths of at least a kilometer. Conditions would have stayed there except for some powerful event that led to a thaw. Perhaps it was volcanic eruptions. Perhaps it was a comet smashing into Earth. The one that least strains probability is volcanoes burning holes in the ice, erupting and injecting carbon dioxide into the cold, dry, rainless air, where it accumulated and started a growing greenhouse effect, warming Earth. Around the equator, parts of ocean lost their white ice blanket, leaving the exposed water to radiate less of the sun's heat away. This started an upward spiral in temperature that ended with a level of atmospheric carbon dioxide nearly 350 times that of today.

The overshoot to extreme greenhouse conditions was pulled back down by the hydrological cycle of evaporation and rainfall. Carbon and bicarbonate were washed from the air and land, and carried into rivers that emptied into the sea. There, it precipitated out of warm sea water along the continental margins, and was deposited as carbonate rock.

At this point, the hypothesis explains the facts of interest that prompted its creation. But as a good hypothesis should, it explains more than its creators put into it. The snowball Earth hypothesis explains why there are now complex forms of life. It goes like this: As Earth returned to life-conducive conditions, microscopic marine organisms that had found refuge in the deep unfrozen part of the ocean, clustering around volcanic vents in the sea floor and possibly at exceptional surface pools of water warmed by hot springs – these microscopic marine organisms ignited a profusion of life forms, the start of the Cambrian age.

Hoffman and Schrag (2000) offer further circumstantial support for the snowball Earth hypothesis. The support is of a biological nature (whereas the hypothesis is of a physical nature). Hoffman and Schrag point out that the great genetic diversity on Earth today supports the hypothesis, saying: "Severe stress encourages a great degree of genetic change in a short time, because organisms that can most quickly alter their genes will have the most opportunities to acquire traits that will help them adapt and proliferate. Hot-spring communities widely separated geographically on the icy surface of the globe would accumulate genetic diversity over millions of years. When two groups that start off the same are isolated from each other long enough under different conditions, chances are that at some point the extent of genetic mutation will produce a new species."

If snowball Earth happened it was likely a one-time event. Why? Because once it happened, ocean worms and sea snails would evolve and colonize the carbon burial grounds in the oceans. The worms and snails would release a steady stream of carbon to the water, and some of it would make its way to the atmosphere. Hoffman remarks: " [Today] we think we have worms and snails to thank. They and many other animals that live on the sea bottom constantly churn muddy sediments searching for bits of food. This contributes to the breakdown of organic forms of carbon and its release into the water, then into the air. . . . Before the advent of [these bottom-dwelling animals], the rocks in Namibia [Africa] show thin layers of undisturbed sediment on the sea bottom. After animals appeared, the fossil sediments were disrupted by feeding trails, borrows, and other signs of grazing activity."

Arguments rarely break out during church services. No one ventures a religious hypothesis counter to church dogma. In science, however, research hypotheses often provoke productive arguments, and the snowball Earth hypothesis has. Four years after the hypothesis was published, Leather et al. (2002) published geological data that goes against it. The data, collected in Oman, suggest that Earth never became completely iced over, but experienced "a number of glacial advances and retreats, reflected in facies variations and relative sea-level cycles in the basin fill." In other words, throughout the epoch that the snowball Earth hypothesis applies to, the hydrological cycle was probably active in places and did not, as the hypothesis has it, completely shut down. Hoffman has noted that this is not a fatal objection (Marzuola 2002), saying that the snowball theory still holds if nearly all of Earth was frozen over.

Whether or not the snowball Earth hypothesis survives scrutiny, it has motivated scientists to devise alternative mechanistic hypotheses. (One of these is called "the global Black Sea hypothesis"; another, "the methane model." See Monastersky (2004) for a discussion of these, and for the points of debate among their various supporters.) Paul F. Hoffman, co-creator of the snowball Earth hypothesis is quoted as saying, "There's a lot of animosity. It's become very personal. That's because it's important." The other co-creator, A. Jay Kaufman, later changed his mind about the snowball Earth hypothesis, and no longer favored it. (The story has more to go. Subsequent to the above researches, the snowball Earth hypothesis was tested with the hypothetico-deductive method, as described in Res. Ex. 4.28).

Let us say three things about this example. First, it demonstrates how mechanistic hypotheses are central to research fields concerned with historical time lines – geology, archaeology, paleontology, and cosmology, to name some. Second, the creation of a mechanism that accounts for a fact is, in effect, the creation of a narrative tale that solves a mystery. You could tell the snowball Earth hypothesis to listeners around a campfire and hold them rapt; it has a plot, a time line, and the listeners would be curious to know "What next?"

Third, the snowball Earth hypothesis epitomizes great science. Yes, it may someday be seen as badly mistaken. New facts will be discovered, so who can be sure it will stick around? No matter, it is great science because it does what great science is suppose to do:

(1) It provides a common explanation for facts that had been puzzling anomalies when considered in isolation. For instance, it accounts for the anomaly of glacial debris capped by carbonate rocks formed from warm seawater.

(2) It explains more than was put into it, providing a plausible explanation of how the Cambrian age, with its eventual profusion and diversity of species, was kicked off.

(3) At the time of its conception, it explained all the facts that the prior hypotheses explained, while it explained facts they did not. For instance, it accounts for geological data collected in Namibia, Africa, something the prior hypotheses could not (Hoffman et al. 1998).

4) It makes scientists itch to criticize it and to conceive a superior theory. For instance, it was the motivating factor in the research of Leather et al. (2002) that suggests that Earth never became completely iced over.

Incidentally, it was conceived by scientists with complementary expertise. One of its team of creators is a field geologist, another a geochemical oceanographer.

3.8 The method of multiple working hypotheses

First we will explain the original method of multiple working hypotheses. Then we will modify the original method, calling the result the modified method of multiple working hypotheses, which makes it better for the way researchers practice today. Finally, we will present the ethics of ensuring that the modified method is practiced.

The original method of multiple working hypotheses. In 1890, the geologist Thomas C. Chamberlin published a paper urging geologists to practice what he called "the method of multiple working hypotheses" instead of practicing what he called "the method of the ruling theory," which historically geologists had practiced. The word "working" means providing a perspective for discovering knowledge. Chamberlin proposed that geologists should keep in mind multiple perspectives – multiple working hypotheses – for interpreting the facts they discover.

Seventy-five years later, Chamberlin's paper was brought to the attention of researchers in all fields when Platt (1964) wrote of it: "This charming paper deserves to be reprinted in some more accessible journal today, where it could be required reading for every graduate student – and for every professor. It seems to me that Chamberlin has hit on the explanation – and the cure – for many of our problems in the sciences. The conflict and exclusion of alternatives that is necessary to sharp inductive inference has been all too often a conflict between men, each with his single Ruling Theory. But whenever each man begins to have multiple working hypotheses, it becomes purely a conflict between ideas."

Taking Platt's suggestion to heart, the journal *Science* republished the paper (Chamberlin 1965). In it, Chamberlin spelled out the shortcoming of the method of the Ruling Theory: "The search for facts, the observation of phenomena and their interpretation, are all dominated by affection for the favored theory [favored hypothesis] until it appears to its author or its advocate to have been overwhelmingly established. The theory then rapidly rises to the ruling position, and investigation, observation, and interpretation are controlled and directed by it. From an unduly favored child, it readily becomes master, and leads its author whithersoever it will. The subsequent history of that mind in respect to that theme is but the progressive dominance of a ruling idea."

Going on, Chamberlin clarified the advantage of research that is guided by multiple working hypotheses, saying, "the effort is to bring up into view every rational explanation of new phenomena, and to develop every tenable hypothesis respecting their cause and history. The investigator thus becomes the parent of a family of hypotheses: and, by his parental relation to all, he is forbidden to fasten his affections unduly upon any one." In time, as evidence mounts, most of the hypotheses will drop out of the picture, leaving the one that seems most probably true.

Why Chamberlin's method is ill suited to much of modern research. In Chamberlin's time, multiple working hypotheses and the background knowledge pertaining to each could be held in a single researcher's mind. This is hardly possible today. Scientists are deep in understanding their particular areas of research, shallow in understanding outside their particular areas. When they come across novel facts that would bear on hypotheses outside their speciality, perhaps they are unlikely to realize the possible significance.

A little story captures this. Suppose two cosmologists are working on competing hypotheses: Sandra on a hypothesis in string theory; Mike on a hypothesis in the theory of noncommutative geometry, an alternative to string theory. Chamberlin appears to them in a dream, and tells each to give as much attention to the other's hypothesis as to their own. They find they can't. Sandra's mind is filled with her special interest, her hypothesis. Mike's mind is filled with his special interest, his hypothesis. Each lacks time and ability for thinking profoundly about the other's hypothesis.

The modified method of multiple working hypotheses. Let's modify Chamberlin's advice that each researcher should practice the method of multiple working hypotheses. The modification: Each *field* of research should practice the method of multiple working hypotheses. With this modified method, the field of research – not the researchers in the field – becomes the parent of a family of working hypotheses. It is a field's duty to ensure that all working hypotheses in the family are fairly addressed. This precept is not adequately met today.

The ethics of ensuring that all working hypotheses are fairly addressed. Back in Chamberlin's time, it took small sums of money to do research. Scientists got funds from their employers – colleges and government institutions – and in some cases funded themselves. Today, when researchers can't be expert in all areas, just their own, research in their areas takes large sums. Those who control the purse strings practice, by and large, the method of the ruling theory. Research into the hypothesis that is favored has the best chance of being funded. Unpopular hypotheses tend not to be adequately funded. In the words of Perry et al. (2004): "Reviews for grants or manuscripts generally involve two or three peer researchers. When you submit a [proposal to research an unpopular] hypothesis, the likelihood of an opposing reviewer reading your proposal is high. If 80% of people disagree with you, the chance of two reviewers agreeing with you is only 4%. In contrast, if your view is accepted by 80%, the likelihood of getting two positive reviews is 64% or 16 times as likely." This system, they conclude, favors the acceptance of grant proposals and the publication of manuscripts that pertain to the one or two ruling hypotheses, while it "stifles opposing views without directly challenging them or disproving them in the scientific arena."

To solve this, an ethic is needed. Let's demonstrate the ethic in the context that there are four main hypotheses of the cause of Alzheimer's disease (Kolata, 1981; Hardy and Selkoe, 2002):

Hypothesis 1: Below-normal levels of acetylcholine is the cause of Alzheimer's. Hypothesis 2: Slow viral infections are the cause of Alzheimer's. Hypothesis 3: Exposure to toxic metals, as aluminum, is the cause of Alzheimer's. Hypothesis 4: Plaques of amyloid β-peptide clogging the brain is the cause of Alzheimer's.

Imagine you are an Alzheimer's researcher receiving funds to investigate the amyloid hypothesis. You'd bet that in fifty years hence research will have established its truth. Imagine further that you are asked to review a researcher's proposal for grant funding into the slow virus infections hypothesis. Ethics demands that you be a reasonable critic. Strictly judge the proposal on the quality of the proposed methodology for reaching a reliable decision on the truth or not of the hypothesis. Show the same professional objectivity when you referee a journal manuscript on the slow virus infections hypothesis. Fairly judge the manuscript on the methodology and arguments the author(s) followed in reaching conclusions.

Every researcher must obey this ethic. Then will scientific progress under the modified method of multiple working hypotheses achieve its full potential.

3.9 Picturing the role of research hypotheses in science

Let's start the picture with facts. Facts, when you come to think of it, are ideas we get through our senses. When we weigh an object, we get the fact, the idea of its weight. When we measure the time of sunrise, we get the fact, the idea of when sunrise occurs.

Facts, whether quantitative or qualitative, have a characteristic property, which Margenau (1961) tells of: "What is factual about a fact is that it is independent of our control: it is simply there; it clamors to be recognized by us as such. A fact is what cannot be denied, what obtrudes itself into the process of knowledge whether we wish it there or not. . . ."

Research hypotheses are ideas too. Unlike facts, whose source is "out there" in nature, we conceive research hypotheses imaginatively. Although they are about the "out there," they don't come to us as facts do from the "out there."

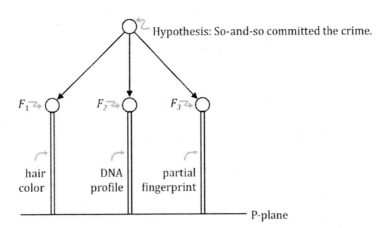

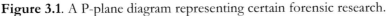

Figure 3.1. A P-plane diagram representing certain forensic research.

Look now at Fig. 3.1, a P-plane diagram representing certain forensic research. In the C-field (the field of constructs) are three constructs tapping into the P-plane, each a double line capped by a circle. The three apply to evidence found at a crime scene: the construct "hair color," for a bit of the apparent murderer's hair; the construct "DNA profile" extracted from drops of blood on a glove worn by the murderer; and the construct "partial fingerprint" taken from a knife used to commit the murder. The three facts, qualitatively measured on these constructs, are denoted by

F_1, F_2, and F_3. The measurement of F_1 is "gray hair," the measurement of F_2 is the particular DNA profile, and the measurement of F_3 is the particular partial fingerprint.

Suppose a suspect we will call "So-and-so" matches the facts: has matching grey hair, DNA profile, and partial fingerprint. Ah! we think, a hypothesis: "So-and-so committed the crime." The hypothesis is shown as a circle in the C-field. If the hypothesis is true, it accounts for each of the three facts. We indicate this by the three arrows coming from the hypothesis to the facts.

The P-plane diagram helps us realize that the larger the number of constructs and their facts that a hypothesis is retroduced from, usually the more we tend to believe the hypothesis is true. For the crime example, if there was just one construct, say that of a partial fingerprint, So-and-so's presumed guilt would be less compelling.

Hypotheses are isolates. **Isolates** are a special type of construct. They are well defined, constructed concepts, but they are isolated from the P-plane (hence their name). They do not directly connect to nature, and so they cannot be measured directly. They are represented in P-plane diagrams as circles connected with arrowed lines to constructs or to other isolates.

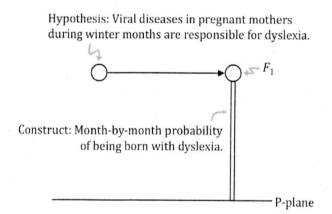

Hypothesis: Viral diseases in pregnant mothers during winter months are responsible for dyslexia.

F_1

Construct: Month-by-month probability of being born with dyslexia.

P-plane

Figure 3.2. A P-plane diagram depicting research on dyslexia that is described in Res. Ex. 3.8.

For another example, look at the P-plane diagram in Fig. 3.2. It could depict the research on dyslexia that is described in Res. Ex. 3.8. The "month-by-month probability of being born with dyslexia" is the construct shown. Suppose it has been measured, and the fact F_1 is that the probability is greatest for children born in May, June, or July. To the left of the construct, shown as a circle, is a hypothesis conceived to account for the fact. The hypothesis is: Viral diseases in pregnant mothers during winter months are responsible for dyslexia. The hypothesis has no direct connection to the P-plane; its connection is indirect, through the construct. The arrow running from it to the construct indicates that the hypothesis deductively predicts the fact F_1 of the construct.

(That the hypothesis in Fig. 3.1 is above the constructs, and the hypothesis in Fig. 3.2 is off to the side of the construct, is meaningless. Usually we will prefer to put the hypothesis off to the side.)

For more on P-plane diagrams, see Margenau's (1961) clear description. To learn more about isolates and their role in research, see Caws's (1957) article. You might find Holton's (1979) article interesting; he describes how Einstein drew P-plane diagrams as an aid to picturing science.

3.10 A list of research examples

To this point in this chapter, the research examples are the start of a view of conceiving and supporting research hypotheses. Most illustrate the fact method (retroductive method); the rest, the question method and the subject method. With the aim of reaching a sweeping view, we continue with more research examples.

Res. Ex. 3.22: Why do ice skates slide easily?

The fact is, ice skates slide easily across ice. Colbeck (1995) offers three possibilities for the main mechanism that explains the fact. Hypothesis 1: The skater's weight is distributed over the small area of a skate's blade. This produces great pressure on the ice, melting the surface, making a thin film of water upon which the skate glides. Hypothesis 2: Friction from rapid movement across the ice melts the surface, producing a thin film of water upon which the skate glides. Hypothesis 3: Ice is covered with microscopic balls of ice, as well as a cushion of water vapor, upon which the skate glides. Naturally, possibly several of these mechanisms are operating simultaneously, with one primary and the others helping, making a fourth hypothesis.

Res. Ex. 3.23: Why did the market for classical literature increase?

The fact is, in Britain it did increase in the early 1990s. Why? A 1992 report in the British trade paper *Bookseller* had this to say: "There are various theories to explain this phenomenon. Among them: the growth in the school and student populations; the introductions of GCSE's [a national student test of knowledge], with their broader syllabuses; the increasing adoption of modular courses at tertiary levels of education; readers going back to tried and trusted works at a time of uncertainty; interest, as the millennium approaches, in the past; escape from the bombardment of modern culture; disillusionment with contemporary fiction." These individual hypotheses are all possible explanations arrived at by retroductive logic. Besides, certain ones may be viewed as contributing causes of the fact, and incorporated into a more inclusive hypothesis.

Res. Ex. 3.24: How was the Milky Way formed?

The story begins with a set of contradictory facts. Reliable measurements of the color and brightness of globular star-clusters indicate that the Milky Way was born 12 to 18 billion years ago. And reliable measurements of the relative abundance of certain radioactive isotopes in meteorites give an age of 9 to 15 billion years. And reliable measurements of the luminosity of white dwarf stars give an age of 7 to 11 billion years.

Conflicting data are always a spur to developing a reconciling explanation, and Mathews and Schramm (1993) did. Their hypothesis: The stars in the Milky Way were born in two waves, separated by billions of years. During the first 12 to 18 billion years, gas clouds coalesced into bunches of newborn stars, including globular clusters. Five to 6 billion years later, gravity rounded up the remaining gas clouds, collapsed them, and the second wave of star birth began.

Res. Ex. 3.25: Why are we prone to gossip?

Even Henry David Thoreau, when he traveled from Walden Pond into town for supplies, took in gossip and recorded some of it in his journals (Krutch 1948). The fact is that to every corner of the globe, practically everybody gossips or listens to it. Why? Hypothesis: The species has a gossip gene.

A gossip gene would have helped bind our ancient forebears into groups (can't gossip without an audience). Relative to loners, group members would have had a larger chance of surviving attacks from people and animals, of recovering from accidental injuries, instilling their genes to their offspring.

Res. Ex. 3.26: Why are more home runs being hit?

The statistical facts of baseball are interesting, and none more so than the rise in the number of home runs. In the major leagues of American baseball, the number of home runs increased from 2,698 hit in 1975 to 5,693 hit in 2000. Gould (2001) reviewed several hypotheses capable of explaining why:

Hypothesis 1: Baseballs have been made more lively. Plausible supporting reason: As more home runs are hit, game attendance rises, bringing more money to team owners, leading them in secret to pressure the official manufacturer of baseballs to make the balls springier.

Hypothesis 2: Pitching is getting worse. It's hard to conceive a plausible supporting reason for this, making the hypothesis more doubtful than hypothesis 1.

Hypothesis 3: Batters are getting stronger. Plausible supporting reason: Players want higher salaries. More home runs lead to higher salaries. Stronger players lead to more home runs. Weight training, taking food supplements, even steroids, lead to stronger players.

A hypothesis that Gould didn't mention is that umpires have gradually shrunk the strike zone to better match "the home run zone." Another that I heard Cal Ripken, Jr. mention is that players have become less embarrassed over striking out, so they hold back less in swinging.

Res. Ex. 3.27: What killed the Athenians?

All their gods and goddesses could not save the Athenians. In the circumstances, the Athenians did what many attacked by microbes do – they died. How do we know? Hear Holden (1996) give an important fact, give alternative hypotheses that are possible explanations of the important fact, and give circumstantial evidence that supports one of the hypotheses:

"According to estimates based on the writings of historian-general Thucydides, [the fact is that] up to 300,000 Athenians – one in every three – were felled during a Spartan siege by a mystery disease whose symptoms included high fevers, blistered skin, bilious vomiting, intestinal ulcerations, and diarrhea. Most victims died about a week after onset. Over the years, people have postulated causes [made hypotheses] ranging from bubonic plague to measles to a combination of influenza and staphylococci – but none seemed to fit.

"But in a letter in the April-June [1996] journal *Emerging Infectious Diseases*, Patrick E. Olson, an epidemiologist at the Naval Medical Center in San Diego, and his colleagues propose that Ebola virus may be the culprit, . . . [noting that] in both the ancient and modern cases, victims died quickly, and both diseases appeared and vanished suddenly. In Athens there were widespread deaths among caregivers, but few among the attacking Spartans. This pattern suggested that the disease, like most strains of Ebola, was spread by personal contact and not by airborne pathogens. But what really got Olson's pulse racing were the reports of 'hiccupping' among 15 percent of the Zaire Ebola patients – an unusual symptom also noted by Thucydides, himself a plague survivor.

"There are some other tantalizing hints: Santorini Island near Athens is the site of a Minoan fresco featuring green monkeys, a species known to harbor Ebola. And because Athens was a major port, the virus could easily have been spread by sailors from ships that traded with Africa.

"Epidemiologist Karl M. Johnson. . . . says that the interpretation proposed by Olson is definitely 'suggestive. . . . You'd have to say it's at least as likely as any of the other ones proposed.' Proving it is going to be difficult, however. The Greeks cremated their dead, along with any genetic evidence from the Ebola virus."

Res. Ex. 3.28: What caused the crash of TWA Flight 800?

Some hypotheses come quickly by nature. A place to see this is the explosion of TWA Flight 800 over Long Island Sound in 1996. Within minutes after the crash, newscasts were centered on three retroduced hypotheses of what brought the plane down. Hypothesis 1: It was a bomb. Hypothesis 2: It was a hostile rocket. Hypothesis 3: It was an equipment malfunction. Other hypotheses came slowly, worked out by those with specialized experience, such as Hypothesis 4: Explosive aviation gasoline was mixed in with the less explosive JP4 fuel that is prescribed for airline use.

A retired USAF Major, in a letter to *The Wall Street Journal* (Templeton 1997), suggested Hypothesis 4 and spelled out support for it, part of which reads: "Somewhere in the world-wide travels of the Flight 800, 747 aviation gasoline must have been substituted for the JP4 either innocently, on purpose, or by mistake. If they did not have enough JP4 to fill the plane, say in Karachi, India, aviation gas from the POL [Petroleum, Oil, Lubricants area which every airport has] could be pumped into the system and the plane would fly just fine. . . . When a scenario like this takes place, there will be a quantity of high-octane fuel locked into the center tank, with its explosive gases, even with constant refilling. This fuel will remain sealed there for months or even years until that moment when either high pressures and temperatures or an electric spark turns that tank into an exploding bomb. . . . There is no other way for a 747 to 'blow up' in the middle unless its center tank contains a quantity, however small, of high-octane aviation fuel and its explosive vapors are ignited."

Res. Ex. 3.29: Who wrote Shakespeare's plays?

With the question method of conceiving hypotheses, literary historians long ago asked: Who wrote a group of great plays including *King Lear*, *Macbeth*, *Richard II*, and *Love's Labours Lost*? We all know what the historians have decided. Hypothesis: William Shakespeare wrote the plays.

In his book, *Is Shakespeare Dead?*, Mark Twain (Twain 1996) examined the circumstantial evidence and reasons the historians cited, added his own, and concluded that the hypothesis had no legs to stand on. From the book, here is a selection of Twain's circumstantial evidence and reason that goes against the hypothesis (the quoted words are his):

Concerning expertise in law: The author of Shakespeare's plays was "limitlessly familiar with the laws, and the law-courts, and law-proceedings, and lawyer-talk, and lawyer-ways." Yet Shakespeare couldn't have had much, if any, training or practice in law.

Concerning literary belongings: Shakespeare died poor and his will mentioned "not a play, not a poem, not an unfinished literary work, not a scrap of manuscript of any kind." Literary people at the time almost always put their literary belongings in their wills, but Shakespeare is silent on such.

Concerning national fame: Shakespeare's death was not a great event. "Nobody came down from London; there were no lamenting poems, no eulogies, no national tears – there was merely silence, and nothing more." Had he written the plays, England would have surely made an enormous to-do over him, at least equal of the to-dos it made when Ben Jonson and Francis Bacon died, but England didn't.

Concerning hometown fame: Shakespeare lived nearly half his life in his native Stratford, "yet when he died nobody there or elsewhere took any notice of it; and for sixty years afterward no townsman remembered to say anything about him or about his life in Stratford." A notable would not have been totally forgotten in the place that produced him.

Overall, of those who attributed the plays to Shakespeare, Mark Twain said that their reasons amount to "guesses, inferences, theories, conjectures – an Eiffel Tower of artificialities rising sky-high from a very flat and very thin foundation of inconsequential facts."

If Twain is right that Shakespeare did not write the plays, who did? Let us mention playwright Bertolt Brecht's idea, which he wrote in his journal in 1940 (Brecht 1993). Hypothesis: A collective wrote the plays. As to the circumstantial evidence supporting this, Brecht said: "What makes me think that a small collective produced Shakespeare's plays is not that I believe a single person could not have written these plays because a single person could not have such poetic talent, be versed in so many fields and have such a broad general education. It is just that technically the plays are put together in a way that leads me to believe I recognise the working methods of a collective. The members of the collective need not always have been the same, it may have been a loose arrangement, Shakespeare may have been the decisive personality, he may just have had occasional collaborators etc, but the leading ideas can equally well have come from some elevated person who constantly used S[hakespeare] as head of script production."

Res. Ex. 3.30: Why do certain species of fish school?

Of the nearly 20,000 species of fish, more than 10,000 are known to swim in schools during at least part of their lives, a fact that Partridge (1982) fills out in detail: "In response to a quick strike [of a barracuda], each fish darts from the center of the school. In such a flash expansion each fish moves radially outward, propelled by a single flick of the tail. The movement resembles a bomb burst. In as little as a fiftieth of a second each fish accelerates from a standing start to a velocity of between 10 and 20 body lengths per second. The entire expansion can take place in half a second. . . . [and] collisions have never been observed."

The hypothesis that Partridge favors: Schooling is an innate defense against predators, and its flash expansion is innately coordinated.

There is logical backing for the hypothesis: Flash expansion confuses predators. By making the decision of which fish to attack uncertain, the chance of making a kill is reduced, aiding the prey species' survival. Hear Partridge: "Because the expansion is created by roughly simultaneous tail flicks throughout the school it seems it cannot be coordinated by any means that would require each fish to register the movements of its neighbors." It must be that each member of the school innately knows where the other members will go in the event of an attack.

Res. Ex. 3.31: Why are snakes' tongues forked?

Very few of us have actually inspected and described the factual condition of snakes' tongues. We leave it to others, trusting their account that the tongues are indeed forked. But why forked? Schwenk (1994) reviewed the chronology of hypotheses, each providing a reason. To begin with, around 350 B.C., Aristotle proposed this hypothesis, writing: "[A forked tongue provides snakes with] a twofold pleasure from savours, their gustatory sensation being as it were doubled." Around 1650, Hodierna proposed this hypothesis: "[Snakes use their forked tongues] for picking the Dirt out of their Noses, which would be apt else to stuff them, since they are always grovelling on the Ground, or in Caverns of the Earth." Another ancient hypothesis: Snakes have forked tongues for catching flies between the tines.

A modern hypothesis that Schwenk retroduced is this: The tongue is an organ that senses chemicals left by passing prey. With a forked tongue a snake can sense a chemical gradient with one flick, estimating the direction in which a prey has traveled past, allowing the snake to decide which way to go in chase.

Res. Ex. 3.32: Why do female adders mate with many males?

Madsen et al. (2002) observed the fact that in meadows of southern Sweden, female adders (snakes of the viper family) engage in multiple copulations with a variety of males. After Madison et al. considered various hypotheses explaining possible species-propagation advantages of this tactic, the one they were able to build the most creditable support for was this:

Hypothesis: Multiple copulating saves the female energy in finding a mate.

Logic supports this. Instead of going about looking for the fittest mate to maximize her reproductive output, it is more energy efficient for the female to collect sperm without judgment from a mix of males. Harboring it in her reproductive tract, when the time comes she lets it loose to compete in a race to fertilize her eggs. This favors her ova being fertilized by sperm from the fittest male, because its sperm are most likely the fastest swimmers.

They say science is always advancing. Partly this means we continually know more and more. And partly it means that new knowledge is calling into question current knowledge, and replacing it. The next research example illustrates an instance where a hypothesis came to be regarded as knowledge and later was disposed of.

Res. Ex. 3.33: What kind of plague caused the Black Death?

The Black Death appeared in southern Italy in 1347, and within three years had spread northward through Europe to Norway. As to a plausible cause of this fact, an early hypothesis that was retroduced and stuck for nearly a century was that bubonic plague caused the Black Death.

Later, new evidence became available: epidemiological records on the outbreak, and medical and biological findings on the pathology of bubonic plague. Drawing from it, Scott and Duncan (2001) discovered that the bubonic plague hypothesis was at odds with 20 items of new evidence. To mention just two: (1) In one place in France the Black Death "jumped more than 30 miles in a few days with no intermediate outbreaks. . . . The speed of this transmission is completely inconsistent with bubonic plague." Assuming flea-carrying rats caused the plague, it could have spread no faster than rats spread, and that's a few miles a year. (2) Flea-carrying rats would rarely infect more than one member per household. Yet there are "records from late medieval Italy showing that 96% of the families examined had multiple cases of plague per household. . . ."

Scott and Duncan retroduced a novel hypothesis that is consistent with all of the items of new circumstantial evidence, epidemiological and historical:

Hypothesis: Hemorrhagic plague caused the Black Death.

Key support for this came from records of autopsies of victims of the Black Death: red spots on the chest, so-called God's Tokens, which isn't a symptom of bubonic plague but is of hemorrhagic plague, a cousin of the Ebola virus. Besides, hemorrhagic plague has features that allow its rapid geographic spread. It has an average 38-day delay from the date of infection to death. During the first 22 days the infected feel normal, and would have gone on with their daily routines, some taking trips of many miles, infecting people they contacted.

There's a lesson here for every field of science. It is that the bubonic plague hypothesis could have been knocked down years earlier. The evidence to do it was available. But researchers assumed the hypothesis was correct because they had always heard people talk of it as if it was correct. As biologist Duncan tells (Guterman 2001): "When the Black Death struck. . . . [people] speedily thought it was an infectious disease. . . . [But around 1900 Alexandre] Yersin discovered bubonic plague. And I think that historians who are completely ignorant of anything to do with medicine or biology said, 'It's a plague, isn't it? Plague – same word? Obviously, it's bubonic plague.' They erected a great superstructure of tales and traditions and they resolutely refused to consider any suggestions to the contrary."

Res. Ex. 3.34: Why do some bird species practice divorce?

Among bird species in which individuals marry, there are times when a partner will leave its mate and re-pair with another. Practicing the method of retroduction to explain this fact, Choudhury (1995) conceived four working hypotheses (and developed support for each that we won't describe):

Hypothesis 1: Partners may be incompatible and not work well as a team. Hypothesis 2: A partner, spying an unattached potential mate that shows signs of better fitness, may leave to join with it. Hypothesis 3: A third bird may intrude itself on a contented marriage, displacing one of the members. Hypothesis 4: A pair may become accidentally separated during migration or bad weather, and re-pair out of necessity.

By publishing these working hypotheses, Choudhury invited ornithologists to judge them and the quality of the support he gave. Making them public opened the way to bring up objections or further points in their favor he had possibly overlooked, to propose further hypotheses, and to devise tests of the hypotheses.

Res. Ex. 3.35: What is the cause of thunder?

We can learn some general points about science by looking at how people, down through the ages, retroduced hypotheses explaining the fact of thunder. To begin with, Aristotle tells in his essay *Meteorologica* (Lee 1952) that Empedocles retroductively argued that thunder is explained by the sun. Hypothesis: Some of the sun's rays get trapped in the upper atmosphere, and then descend to where they flash as fire through the clouds they meet. Lightning is the fire, and thunder the noise of its being quenched.

Later in the essay, Aristotle retroduces his hypothesis, which has nothing to do with the sun. Hypothesis: Thunder is caused by "the dry exhalation that gets trapped when the air [in clouds] is in process of cooling [and] is forcibly ejected as the clouds condense and in its course strikes the surrounding clouds, and the noise caused by the impact is what we call thunder. . . . [That is,] the windy exhalation in the clouds produces thunder when it strikes a dense cloud formation." Going on, Aristotle tackles the details of thunder: "Different kinds of sound are produced because of the lack of uniformity in the clouds and because hollows occur where their density is not continuous. . . . As a rule, the ejected wind burns with a fine and gentle fire, and it is then what we call lightning, which occurs when the falling wind appears to us as it were coloured. Lightning is produced after the impact and so later than thunder, but appears to us to precede it because we see the flash before we hear the noise."

Nearly 2,000 years after Aristotle, René Descartes conceived a different mechanism. Hypothesis: When higher storm clouds lose their buoyancy, they fall and strike clouds below them. This breaks the ice in the clouds, causing the sparks and noise that are lightning and thunder.

Thomas Hobbes adapted Descartes' hypothesis, keeping the thunder-making role for ice but changing the rest. Let us quote from Hobbes' philosophical papers (Molesworth 1845). Hypothesis: "You know that it is only in summer, and in hot weather, that it thunders; or if in winter, it is taken for a prodigy. You know also, that of clouds, some are higher, some lower, and many in number. . . . with spaces between them. Therefore, as in all currents of water, the water is there swiftest where it is straitened with islands, so must the current of air made by the annual motion be swiftest there, where it is checked with many clouds, through which it must, as it were, be strained, and leave behind it many small particles of earth always in it, and in hot weather more than ordinary. . . . The particles are enclosed in small caverns of the ice; and their natural motion being the same which we have ascribed to the globe of the earth, requires a sufficient space to move in. But when it is imprisoned in a less room than that, then a great part of the ice breaks: and this is the thunder-clap. The murmur following is from the settling of the air. The lightning is the fancy made by the recoiling of the air against the eye."

Two general points are apparent from the above:

General point 1: Mechanistic hypotheses, as those above, are deductive. They are built from a number of "if-then" ideas, although equivalent words and expressions for "if" and "then" are commonly used in stating them. To illustrate, here is how Descartes' deductive chain runs with if-then statements: If higher storm clouds lose their buoyancy, then they fall. If they fall, then many of them strike clouds below. If they strike clouds below, then the ice in them breaks. If the ice breaks, then lightning and thunder are produced.

No doubt about it, early philosophers were expert in the use of if-then deductive logic, and they were driven to apply it. Reincarnated to the present and given today's if-then ideas to think with, they would shine as good theorists. It is just that the if-then ideas they had available were, although they did not know it, mostly silly, based on false analogies from common experience. It is true, for instance, that if a kettledrum is dropped from a great height onto another kettledrum, then there will be sparks and noise. But carrying this over by analogy to say that if a cloud drops onto another cloud, then there will be sparks and noise, is incorrect.

General point 2: Answers to questions can be sought and found on different conceptual levels, each with constructs and isolates appropriate to it. The above philosophers retroduced possible explanations of thunder in terms of the conceptual level of clouds in the sky and familiar concepts of sun, wind, and ice. By the early 1900s, scientists had retroduced mechanistic hypotheses involving physical properties at a level our senses cannot perceive (Few 1975). Hypothesis 1: A stroke of lightning creates a vacuum, and thunder is the noise of air rushing in to fill it. Hypothesis 2: A stroke of lightning turns water droplets in its path into an explosion of steam. Hypothesis 3: A stroke of lightning separates water molecules by electrolysis into hydrogen and oxygen, which recombine explosively. Hypothesis 4: A stroke of lightning heats the air in its path, and the heated air causes a shock wave in the surrounding air, which decays into an acoustic wave.

By 1960, technology had advanced to the point of allowing laboratory experiments. Under controlled conditions, high voltage discharges of current could be let loose through both moist and dry air, and shock waves and their decay measured. From experiments, researchers decided against the first three hypotheses above and confirmed the fourth.

Res. Ex. 3.36: What is behind those oddly arranged newt tracks?

The Coconino sandstone layer in the Grand Canyon, it's long been known, formed as the solidified remains of sand dunes from the Coconino period. Brand and Tang (1991), after discovering an important fact about fossilized tracks of newts in the sandstone, retroduced a hypothesis capable of explaining the fact:

Fact: Fossilized tracks of newts in the sandstone show that the newts had been moving in one direction while their feet pointed in a different direction. Hypothesis: Small streams must have been running through the dunes during the Coconino period. Support: Walking on dry land, newts couldn't have made tracks like these. They must have been walking underwater, pushed sideways by a current.

Res. Ex. 3.37: Where do tektites come from?

O'Keefe and Glass (1985) provide two retroduced hypotheses able to explain the fact that tektites exist on Earth (the numerals in parentheses denote supporting references in the report): "Tektites, unlike stony or iron meteorites, cannot originate outside the earth-moon system because they lack the isotopic indications – that is, adequate levels of ^{10}Be, ^{26}Al, and so on – of exposure to primary cosmic rays over periods of 10^6 to 10^7 years (1,2). From their distribution on the earth, it is clear that, whether terrestrial or lunar, they were launched by a powerful mechanism, presumably either volcanism or some kind of impact event. Earth volcanism is too feeble to produce the observed strewn fields of tektites, up to halfway around the earth (3), and impact on the moon would yield objects with much the same composition (anorthositic gabbro or basalt) as most of the lunar crust. We are thus left with two alternatives for their origin: namely, meteorite impact on the earth or volcanic ejection from the moon."

Res. Ex. 3.38: Why does the smallest frog species lay few eggs?

From Vergano (1996) comes a summary of the findings and conclusions of two biologists, S. B. Hedges and A. R. Estrada:

Fact: Unlike species of larger frogs, which lay many eggs, depositing them in ponds and streams, the smallest frog species known – an adult would fit on a thumbnail – lays one egg at a time, on land.

Hedges and Estrada retroduced a hypothesis capable of explaining why: Conserving energy is less a concern with larger frog species. With them, the period from egg to frog is practically nonstop predation on their tadpole larvae. A profusion of eggs offers insurance that enough tadpoles make it through. With the smaller species, it would tax their energy to lay so many doomed eggs. A strategy that skips the water, "leapfrogging" predation, is one solution.

Res. Ex. 3.39: Why do whales leap?

In *Moby-Dick* Herman Melville wrote a literary form of the fact that whales leap: "Rising with his utmost velocity from the furthest depths, the Sperm Whale thus booms his entire bulk into the pure element of air, and piling up a mountain of dazzling foam, shows his place to the distance of seven miles and more. In those moments, the torn, enraged waves he shakes off, seem his mane"(Whitehead 1985). What is responsible for the fact of the leaping? Whitehead reviewed three possibilities:

Hypothesis 1: Leaping whales are engaging in play. Support: With no notable exceptions, mammals play, and so apparently need to.

Hypothesis 2: Leaping whales are communicating by sound. Support: Whales need a foolproof way of communicating. There is a risk of getting separated and lost from one another at times when the noise of wind and waves masks their normal vocalizations. The slap of their reentering the water is not easily drowned out.

Hypothesis 3: Whales leap in an act of courtship, displaying their power. Support: From people to butting rams, courting animals commonly attempt to impress with displays.

Res. Ex. 3.40: What explains the Exodus?

The Old Testament Bible reports the reputed fact that the Red Sea parted, exposing the sea floor, allowing the Israelites to walk across. Once across, the sea filled the gap, drowning the pursing Egyptians. Nof and Paldor (1992) retroduced a possible physical explanation. Hypothesis: Strong and constant winds bared the sea floor. A plausible supporting reason: It is physically possible that winds of 40 knots or more, lasting 10 or more hours, could have blown a mass of water up on shore, dropping the sea level by 10 feet, allowing the Israelites to cross. With the pursing Egyptians in the gap, if the wind stopped the gap would have filled in. This doesn't necessarily dismiss a religious interpretation. It provides a physical way for divine intervention to do the job by causing a great wind.

Res. Ex. 3.41: Why did the students suddenly feel ill?

On April 13, 1989, as 2,000 spectators and 600 student performers, 6th through 12th grade, gathered in Santa Monica's Civic Auditorium for the annual Stairway-of-the-Stars concert, an interesting fact occurred. Headache, dizziness, weakness, abdominal pain, and nausea spread among groups of students. Sixteen of the soprano girls fainted. Eventually 247 students fell ill. The auditorium was evacuated. No evidence of fumes or toxic materials was found.

From studying firsthand accounts provided by spectators, and from surveying the student performers with a questionnaire, Small et al. (1991) proposed a cause. Hypothesis: It was mass hysteria. They offered a variety of support: The symptoms of mass hysteria were abundant, especially among girls. It appeared to have spread through social networks, by sight, sound, or both. The best predictor of developing the symptoms was observing a friend with the symptoms. The ill tended to hyperventilate; some experienced psychological stress at the thought of having to perform; there was no illness among the spectators or those running the show. Similar outbreaks have occurred elsewhere in the past; in U.S. schools, one or two outbreaks resembling mass hysteria have been noted per year.

The research has practical application. It brings to the attention of school and health officials, and police and firefighters, a diagnostic symptom for deciding whether a mass illness has a physical or a psychological cause.

Res. Ex. 3.42: On what day of the year are you most likely to die?

When a number of independent retroductions all suggest the same sort of hypothesis, very probably there is something to it. Consider:

Fact: It is common knowledge that of the 36 dead U.S. Presidents, three – John Adams, Thomas Jefferson, and James Monroe – died on July 4, a significant day to them (Martin 1998). Retroduced hypothesis: Solidly patriotic and fated to die as July 4 approached, they willed their reaching it and not passing it by.

Fact: In a sample of 1,919 Jews, their mortality dropped to 31 percent below normal as Passover neared, and rose afterward. Retroduced hypothesis: Many of the dying wanted to reach this significant religious day. (Phillips and King 1988).

Fact: In a sample of 1,288 Chinese, the death rate dipped approaching a holiday and rose just after (Phillips and Smith 1990). Retroduced hypothesis: Holidays are unusually significant times for Chinese. Some Chinese, feeling they would die near a holiday, managed to hang on, letting go just after the holiday.

Fact: From a sample of 2,745,149 adults, Phillips et al. (1992) found that women were more likely to die in the week following their birthday than in any other week of the year. They retroduced a hypothesis explaining why. In their words, "females are able to prolong life briefly until they have reached a positive, symbolically meaningful occasion."

Fact: From the same sample of 2,745,149 adults, Phillips et al. (1992) found that the mortality of males peaked shortly before their birthdays. Retroduced hypothesis: On balance for males, their birthdays are a negative event, serving "as a signal that one promising period of life is over and another less promising one is about to begin." Some males may examine their lives as their birthdays near and get depressed from perceiving a shortcoming between their aspirations and their achievements.

A practical consequence of their and related research, believe Phillips et al. (1992), is that surgeons may want to avoid scheduling elective operations that coincide with symbolically meaningful occasions.

Res. Ex. 3.43: How did insect wings originate?

The question method of conceiving hypotheses sometimes starts from the meeting of several lines of common sense. Take insect wings. Common sense tells us that the wings of flying insects have to be large enough to lift and propel the insects. And common sense tells us that a mutation in wingless, ground-based insects would have been unlikely to produce wings big enough to give the necessary lift. Question: How, in light of this, did insect wings originate?

Marden and Kramer (1994) saw a clue in flightless stoneflies that have stubs which help them skim across the surface of water. Their hypothesis: A mutation on a stonefly brood gave a larger stub that helped lift some stoneflies briefly into the air, giving them an extra measure for surviving. A new species of flying stoneflies sprang from the old (which continued to exist too). After this revolutionary mutation occurred, evolutionary mutations followed, refining the wings and their use, bringing their size above the threshold for sustained flight.

Res. Ex. 3.44: Why all those cattle bones in the barrow (burial mound)?

Davis and Payne (1993) report on the remains of a man who died 4000 years ago in Irthlingborough, England. The goods accompanying the man in his burial were of unusually fine quality, a sign he was wealthy. Near the man and the goods were skulls, mandibles, scapulae and pelvises of about 185 domestic cattle. Many of the skulls had missing incisors, indicating a delay of a month or more between slaughter and final placement of the bones in the grave (soon after death, incisors begin falling out). Quoting Davis and Payne: "The 185 cattle would have provided at least 40 tons of meat which, on a ration of 1 kg per person per day, would feed 40,000 people for a day, or 500 people for two-and-a-half months. . . . [The barrow at] Durrington Walls in Wiltshire, which covers 30 acres, might have required between 50,000 and 90,000 man-days to construct. For the far smaller barrow at Irthlingborough, 185 cattle would be vastly more than necessary just to feed the builders during its construction."

To explain these facts, Davis and Payne hypothesized that the cattle bones were there because the man was special, and were not leftovers of cattle slaughtered to feed the people who built the burial mound.

For support of their hypothesis, they drew upon ethnographic considerations, citing parts of Madagascar today where the dead person is put in a temporary burial place, and several months later reburied with a large number of cattle bones.

Finally, Davis and Payne thought it reasonable to generalize their hypothesis beyond this man, to the view that ancient Britons used cattle as an integral part of the funeral rites of the wealthy.

Res. Ex. 3.45: Have the French their own ecology of eating?

Even a casual survey of the subject of obesity will turn up the fact that Americans are, on average, overweight relative to the French. But casual is not good enough for science. Rozin et al. (2003) decided to be scientific. For France and for America, they sampled the sizes of individual food portions stated in cookbooks, the sizes of pre-packaged individual portions of food and nonfood products in supermarkets, the time spent eating, and the amount of snacking. From this they estimated that the French eat fewer calories than Americans do, and that the French are less obese. To explain the fact, they proposed a hypothesis. It has to do with the French culture, the unhurried pace of life, which affects what they call "the ecology of eating" (for details, see their article).

4
How to test research hypotheses

Right at this moment you are about to test the following hypothesis: "Earth will stop rotating in five seconds."

One, two, three, four. . . . , time's up. Were you launched eastward at hundreds of miles per hour? Yes? – then that corroborates the hypothesis. No? – then that disproves it.

This demonstrates the hypothetico-deductive method of testing hypotheses. For short, it is also called the H-D method, and tests of hypotheses with it are called H-D tests. It has been practiced for hundreds of years, but not until the middle of the 20th century did immunologist Peter Medawar gave it its name, which captures its aspect of deducing testable predictions from hypotheses.

To gain a more definite idea of it, let's go though the demonstration again. First, we state the hypothesis:

H: Earth will stop rotating in five seconds.

From *H* we deduce a test prediction, *P*, i.e., a predicted fact.

P: We'll all be launched eastward at hundreds of miles per hour.

The deduction of *P* from *H* means that if *H* is true, then *P* must be true. This, of course, doesn't mean *H* is true, for whether it is or not is what we are trying to find out. It means that in the event *H* is true, *P* has to be true too.

Now comes the rest of the hypothetico-deductive method. Nature gives us the actual fact, *F*. Suppose it is that we are launched eastward at hundreds of miles per hour. Accordingly, *P* = *F*, i.e., the predicted launching is true to the actual launching. This corroborates the hypothesis ("corroborates" means "makes us more willing to wager that the hypothesis is true"). It doesn't prove the hypothesis is true: a west-to-east comet hitting Earth could have been what launched us.

Very well, but suppose instead that the fact, *F*, is different. Five seconds pass and we are not launched. This disproves the hypothesis (and, besides, rules out a comet).

Tyson (1989) mentions another way to do a hypothetico-deductive test of *H*. From *H*, he deduces a different test prediction:

P: The Pacific Ocean will wash up onto North and South America, and the Atlantic Ocean will wash up onto Europe and Africa.

The point to bear in mind is that there is more than one way to test a given research hypothesis.

Why not dispense with test predictions deduced from the hypothesis? Instead, why not test the hypothesis directly by having an astronaut in stationary orbit watch Earth to see if it stops rotating? Yes, in this instance we could do a direct test. Yet direct tests in scientific research are seldom possible. Typically, research hypotheses are isolates (as explained in section 2.17). We cannot look to see by inspection whether isolates are true or not. The way around this is to test hypotheses indirectly, through their test predictions.

There you have it – the core of the H-D method. In this chapter I would like to show you how to apply it to gain reliable knowledge.

Section 4.1 illustrates the application of the H-D method in terms of two everyday examples, one about a car, one about a dam on a river. Section 4.2 formalizes the H-D method as a series of steps. Section 4.3 takes you through a tour of diverse research examples based on the H-D method. Section 4.4, on performing multiple H-D tests of a given research hypothesis, and section 4.5, on using control groups in H-D tests, show how these two approaches excel at wringing out uncertainty about the truth of research hypotheses. (When I was in college it would have been good to know the material in section 4.5. In effect, several professors had independently said, "Hypotheses can never be proved true. They can only be proved false." I asked one professor why and he said he didn't know, but had been told so! As section 4.5 shows, a hypothesis that is correct can in many instances be absolutely proved to be correct.)

Going on, section 4.6 explains three ways of testing hypotheses economically with the H-D method. Section 4.7 explains how progress in science depends on both the retroductive method and the H-D method. Section 4.8 explains why researchers have to be good at using both the retroductive method and the H-D method. Section 4.9 explains how to present H-D tests in journal articles and seminar talks. Finally, section 4.10 explains two meanings of the word "theory."

To close these opening remarks, here's a thought for spotting a key difference between retroduction (covered in chapter 3) and hypothetico-deduction (covered in this chapter). With retroduction, it's "fact first, hypothesis last": first a fact is obtained, and then a hypothesis is conceived that is able to account for the fact. With hypothetico-deduction, it's "hypothesis first, fact last": first a hypothesis is conceived, then from the hypothesis a predicted fact is deduced, and then the (actual) fact is obtained.

The H-D method doesn't depend on how the hypotheses tested with it are conceived. They may be conceived with the retroductive method (i.e., fact method), the question method, the subject method, or even be discovered by accident.

4.1 Everyday examples of the H-D method

The following two everyday examples make a nice introduction to the H-D method. The first is about a car, the second about a dam on a river. In explaining them and the other examples in this chapter, we will use the following symbols, some set in italic type, and some in regular type. H stands for the hypothesis being tested. P stands for the test prediction, i.e., the fact predicted from the hypothesis. F stands for the actual fact from an experiment. When we discuss different possible facts from an experiment, we will subscript F as F_1, F_2, F_3, etc. And last, T stands for true, and F stands for false.

The car example. Suppose a car has been parked outside your home for three days, and an unsigned note on its windshield reads, "Out of gas." But is it really out of gas? The car's gas gauge is broken; the cap on its gas tank is locked. So you decide to do a H-D test of the hypothesis:

H: The car is out of gas (i.e., its tank is empty of volatile liquid fuel of any sort: gasoline etc.).

To do the test, you assume the hypothesis is true, i.e., that the car is really out of gas. On this assumption, you deduce a test prediction, P, i.e., a prediction of a fact. There are several you could deduce. Suppose you go with this one:

P: The car will not start.

In other words, you use if-then logic, another name for deductive logic. You think, "If it is true that (H) the car is out of gas, then it is true that (P) the car will not start." To save words, we let the phrase "it is true" be implied, reducing what you think to: "If (H) the car is out of gas, then (P) the car will not start." This "if H, then P" idea, you believe, is true.

Next, you do an experiment and get the fact F that corresponds to the prediction P. You jump the ignition switch and try to start the car. Suppose the fact is this:

F_1: Grind, grind, grind – the car doesn't start.

Thus the test prediction, P, is true, i.e., $P = T$. This is consistent with the hypothesis being true. However, it doesn't guarantee it is true. The hypothesis could be false. The car could have gas in its tank while something else keeps it from starting – possibly a broken distributor wire, fouled spark plugs, a clogged fuel filter, or a blown head gasket – things we will call "shunt causes."

For instance, a broken distributor wire could be causing the test prediction – "the car will not start" – to be true even if the hypothesis – "the car is out of gas" – is false, and really the car has gasoline in its tank. (At the beginning of this chapter, for the hypothesis "Earth has stopped rotating," a west-to-east comet hitting earth is a shunt cause, as that could make the prediction of our being launched eastward at hundreds of miles per hour be true even if the hypothesis is false.)

In general, a **shunt cause** is any thing that may cause the test prediction, P, to be true to the fact, F, when the hypothesis H is actually false. There are two kinds of shunt causes: known shunt causes and unknown shunt causes. Known shunt causes are those we know are possibilities, e.g., a broken distributor wire. Unknown shunt causes are those we do not know are possibilities. Being beyond our knowledge, we cannot give examples of them.

Back to the car. What do you conclude from the fact that it doesn't start? You conclude that the hypothesis that the car is out of gas is corroborated. The hypothesis passed the test, which raises your belief in it a notch. But it doesn't raise it to 100 percent, because a shunt cause – a known one or unknown one – may possibly be operating.

Now, suppose instead the experiment gives this fact:

F_2: Grind, varoom, varoom. The car starts.

In this case, the test prediction is false, i.e., $P = F$. Consequently, the hypothesis is false. Does that mean definitely false, definitely disproved? Yes it does, provided the "If H, then P idea" is true, and provided the fact is not in error.

Concerning the truth of the if-then idea, it could be that the idea "If (H) the car is out of gas, then (P) the car will not start," which we believe is true, is actually false. In other words, from H we made an erroneous deduction of P. How could this happen? Perhaps, though we don't know it, there is some residual gas in the fuel line, enough to start the car, even though its tank is empty, out of gas.

Concerning experimental error, it could happen that the car didn't start and the "varoom, varoom" noises were made by another car nearby, but we didn't notice this. In this example, the possibility is fanciful, but in nearly all cases of research it isn't. A research experiment can, unknown to the researcher, contain considerable random error and/or systematic error. The researcher takes the experimental fact to be correct when it is actually erroneous. A test prediction that comes out to be true or false to the fact of an erroneous experiment says nothing definite about the truth or falsity of the hypothesis.

This is a convenient place to introduce three best research practices in applying the H-D method:

(1) Methodically search your mind for every condition that could make the "If H, then P" idea, which you take to be true, be actually false. For such conditions that conceivably could be present, make sure they aren't. In the case of the car, blow compressed air through the fuel line before you try to start the car, ridding it of any residual gasoline. Don't go on with the H-D test until you believe that the "If H, then P" idea is really true.

(2) Methodically identify every known shunt cause. Clamp down on those you can control, so they can't operate. Specifically, verify that none of the distributor wires are broken, that none of the spark plugs are fouled, that the fuel filter isn't clogged, that the compression in the cylinders isn't low, and so on. For any shunt cause you find to be operating, control it so it isn't, e.g., fix the broken distributor wire, clean the spark plugs, etc. Then if it turns out that the test prediction is true (because the car doesn't start), that will strongly support the hypothesis that the car is out of gasoline.

Still, you can never be absolutely sure that some unknown shunt is not operating, making the test prediction be true. Because you aren't aware of the possibilities of such shunt causes, it never strikes you to try to clamp down on them, so they cannot operate. A case would be an enormous solar flare on the sun playing havoc with the car's electronic distributor, something you'd probably never consider. (If you did consider it, you could clamp down on it by waiting until the flare died out before doing the test.)

(3) Design the experiment to make the facts from it be tolerably free of random and systematic error. In the case of the car, should you hear "varoom, varoom," be sure the sounds are coming from the car you are testing, not from some other car.

When you believe you have met these three best research practices, what then? Run the experiment. Should P agree with the experimental fact, that means H is strongly corroborated, meaning quite probably true. And should P not agree with the fact, that means H is quite probably false. We say "quite probably" instead of "definitely" because we cannot be absolutely certain we have actually achieved all of the above three practices.

Fig. 4.1 shows a P-plane diagram for testing a hypothesis with the H-D method. The hypothesis, H, is an isolate, and so is not directly attached to the P-plane. The arrow from H to P indicates the deduction of the test prediction, P, from H. The arrow reads "If H, then P." The short slanted line on the arrow stands for all shunt causes, known or unknown, which could make the prediction P be true in the event the hypothesis H is false. The two parallel lines coming up out of the P-plane are for the construct on which we measure the experimental fact, F. We compare P to F, which tells us whether P is true or false. For the car example, the construct is "result of trying to start the car," and the measured fact is either "car starts" or "car doesn't start."

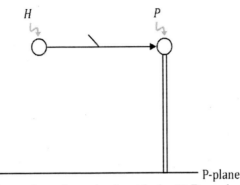

Figure 4.1. A P-plane diagram for testing a hypothesis with the H-D method.

Comment: It is natural to wonder why we don't make the hypothesis, H, be "the car is not out of gas," and make the test prediction be "the car will start." The answer is that the "If H, then P" idea isn't true. Not every car with gas in its tank will start.

Comment: I mentioned above that other predictions, P, could be deduced from the hypothesis, H, that the car is out of gas. That is, P doesn't have to be "the car will not start." An alternative P is "rocking the car will reveal no sloshing sounds in the gas tank." Another is "tapping on the tank will give a hollow ring."

A given test prediction requires a given experiment to see if it is true or false. In scientific research, because some experiments cost more than others, economics is a consideration in choosing what test prediction to use. And as some experiments have more random error and/or systematic error than others, that is another consideration in choosing what test prediction to use.

The dam-river example. Picture, as diagramed in Fig. 4.2, a river in a remote area. Call it the main river. It has a dam on it, and 20 miles downstream from the dam, at place A, is a water-level gauge connected to a radio transmitter. Once a minute it sends a signal to a device you carry around, telling you how high the water level is. In case of a flood, the device buzzes. Notice, too, that between the dam and the gauge, another river – call it the shunt river – flows into the main river.

One day when you are in Chicago, your phone rings. The caller blurts out, "The dam has just broken," then hangs up. It could be a crank call or it could be serious. You propose this hypothesis:

H: The dam has just broken.

Is this really true? You are unsure. Not being at the dam, you can't look at it and know directly. So you decide to test H indirectly, i.e., with the H-D method. To do this, from H you deduce a test prediction, P. You think, "Suppose it is really true that the dam has broken. What test prediction must then also be true?" You come up with this one:

P: The device will soon buzz.

Now it is nail-biting time. What will the fact be? Will the test prediction, P, match the fact or not? Suppose nature soon delivers this fact:

F_1: The device buzzes.

This makes the test prediction true, i.e., $P = T$. The predicted fact was that the device would buzz, and the actual fact was that it did buzz. Hence, the hypothesis is corroborated, i.e., is probably true.

Now let's back up. Suppose the actual fact is instead this:

F_2: The device does not buzz.

This makes the test prediction false, i.e., $P = F$. Therefore, the hypothesis must be false, keeping in mind that this means it is "quite probably false."

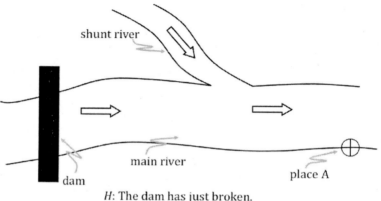

H: The dam has just broken.
P: The device at place A will soon buzz.

Figure 4.2. A section of river, for illustrating a H-D test of the hypothesis (H) "The dam has just broken."

From this example (and the one above about the car), you can see that a good many things can go wrong in testing hypotheses, possibly leading to erroneous conclusions.

Some of the things that could go wrong:

(1) You might fail to notice that you have made a logical error in deducing the test prediction. Perhaps the test prediction, which you believe follows logically from the hypothesis, really doesn't follow. The "If H, then P" idea – "If the dam has just broken, then the device will soon buzz" – may be false while you believe it is true. This could happen from events you do not know about, e.g., an earthquake causing the dam to break and at the same time producing a landslide into the river below the dam, creating a lake, so that the flood from the broken dam cannot reach the water-level gauge. Or perhaps the reservoir behind the dam was recently drained but you weren't told. Either case, landslide or drained reservoir, would make the deduction of the predicted flood from the hypothesis be wrong.

(2) Suppose the "If H, then P" idea is a valid deduction. Still, you might fail to realize that a shunt cause may be operating. Just conceivably, the dam does not break but a cloudburst in the watershed up the shunt river causes a flood that makes the device buzz.

(3) Suppose the "If H, then P" idea is a valid deduction. And suppose no shunt causes are

operating. Yet, you might fail to notice that an error has messed up the experimental fact. The battery in the device may be dead, so it cannot buzz in the event of a flood. Or the battery may be good but the device malfunctions, buzzing when it should not, a false alarm.

One way to minimize the effects of shunt causes is to use a test prediction that admits few known shunt causes. Here's the idea: When we use the prediction, P, "The device will soon buzz," we cannot control the shunt cause of a cloudburst on the shunt river. It's impractical, for instance, to install a dam on the shunt river to catch the waters from a cloudburst. But look here. There is a test prediction (actually a number of them) that isn't subject to any known shunt causes. To test the hypothesis that the dam just broke, use this test prediction:

P: A water pressure gauge at the bottom of the reservoir in back of the dam will register a sudden drop in pressure, making a device we carry around buzz.

It is hard to imagine any shunt cause that can make the water pressure drop while the dam is standing. If the fact is that the device buzzes (and we believe it is not malfunctioning), we will be quite certain the dam has broken. We will be more certain of this than had we used the test prediction involving the measurement of the water level downstream.

In Fig. 4.2, notice that the physical layout of this example – a main river with a shunt river – echos the logic of the hypothetico-deductive method. Understand this example and the car example, and you will understand much of the H-D method as it is used in research.

4.2 The five steps of testing any research hypothesis

We are going to formalize the process of testing research hypotheses, breaking it into five steps. In explaining the five steps, we will assume the hypothesis is a qualitative hypothesis, i.e., expressed in words. After we finish explaining the five steps, we will make a change to step 2 for when the hypothesis is quantitative, i.e., expressed in mathematics.

The five steps are these:

1. The hypothesis step: Conceive a hypothesis, H, by any of the three methods discussed in chapter 3: the retroductive method (i.e., the fact method), the question method, or the subject method. Or, take an existing hypothesis that has been proposed by other researchers. Once you have the hypothesis, follow the advice in chapter 3 of supporting it with circumstantial evidence and with reasons. Why support it when you are going to go on to test it with the H-D method? It's important that the hypothesis you'll test has some likelihood of being true, and obviously the more the better. In point of fact, scientists don't go around testing hypotheses they already suspect are false. The hypothesis that there is life deep in the sun is so certainly false that no sensible person would try to devise a H-D test of it. (The next four steps make up the H-D method.)

2. The deduction step: At this step, conceive a true "If H, then P" idea. That is, from the hypothesis, H, deduce a test prediction, P – a predicted fact of a construct. For qualitative hypotheses, the deduction is done by reasoning with words, as in the above car example and dam-river example.

One way of deducing P from H is to say to yourself, "Suppose, for a moment, that the hypothesis, H, is true. Under that condition, what predicted fact of what construct must also be true?" The answer to this is a test prediction, P. Another way, closely related to this, is to say to yourself, "Suppose I believe that the hypothesis, H, is true. Then what else, what fact, must I also have to believe is the case?" Your answer constitutes a test prediction, P.

3. The shunt cause step: Methodically hunt for known shunt causes that are controllable. You don't want them making mischief, causing the test prediction, P, to be true, even though the hypothesis, H, might be false. Design your experiment to clamp down on them so they cannot operate. For intermittently operating shunt causes, as could happen with observational experiments in the field, monitor them to verify that they are not operating while you are running the experiment. Furthermore, check for the possibility of having an experimental control group (covered in section 4.5) as a way of rendering ineffective the operation of unknown shunt causes.

4. The experiment step: Design your experiment in a way that minimizes the random error and the systematic error in it. Then run the experiment and learn the fact against which the test prediction will be compared.

5. The decision step: Compare the test prediction, P (from step 2), to the actual fact, F (from step 4). If P agrees with F, to within a margin of error you believe is acceptable to researchers in your field, that corroborates the hypothesis. If instead the test prediction, P, does not agree with the fact, F, that falsifies the hypothesis.

Comment: To test a hypothesis, H, by deducing several different test predictions, and then comparing them to the facts of several experiments (as covered in section 4.4 which follows this one), cycle through steps 2, 3, 4, and 5 separately for each experiment.

Comment: For mathematical hypotheses, say a mathematical theory, you must mathematically deduce P from H. To do this, write down on paper the mathematical statement of the hypothesis, H. Transform this statement mathematically, deducing a further mathematical statement. If necessary, mathematically transform this further statement, deducing a further mathematical statement from it, and so on. By this you make a mathematical chain of deductions, artfully and intuitively guided so that at some point in the chain, the mathematical statement deduced makes a test prediction, P, to compare to a quantitative fact from an experiment. To mention just one instance, from a mathematical version of what is known as Errington's threshold-of-security hypothesis, which concerns the number of individuals of certain species of wildlife that survive from year to year, I mathematically derived a test prediction (Romesburg 1981).

An illustration of applying the five steps. I would like to sketch the design of a research project that follows the five steps to test a qualitative hypothesis. A real research project to test the hypothesis would take hundreds of hours to conceive and execute, with the payoff being that the test would be more reliable. Still, the sketch is instructive:

(1) The hypothesis step: Suppose we get the hypothesis by the subject method. The subject we had been thinking about is the theory of evolution, and one day we wake up with the following hypothesis in mind:

H: People are innately more comfortable in seeing profile views of dangerous wild animals than in seeing front-on views.

Next, we gather support for the hypothesis. One line of support is an argument based on the greater survival value of profile views to our ancient ancestors. Animals are easier to hunt when observed in profile, as they present a larger target. And animals are less a threat when observed in profile, as they are not in the best position to notice or attack us. For these reasons, it is conceivable that when our ancient ancestors saw a lion or any large wild animal that did not see them, they moved around to the animal's side, avoiding facing it front on. This raised their chance of living and passing on their genes that favored viewing animals in profile. These genes presumably would remain active in us today.

(2) The deduction step: We deduce a test prediction. We think, "If *H* is true, what test prediction *P* – what predicted fact of what construct – must also be true?" Any number of test predictions might be deduced. (The best would be least susceptible to being made true by shunt causes.) Suppose we settle on the following test prediction, which involves a group of randomly chosen people (research subjects), none with experience of being around dangerous wild animals. The subjects should be nearly the same age, half males, half females. We'll expose each to pictures in which dangerous wild animals appear unexpectedly, sometimes in front-on views, and sometimes in profile views. The test prediction is this:

P: On average, when front-on views of dangerous animals are presented, the research subjects' facial muscles, as scored by the Facial Action Coding System (FACS), will show a greater fear response, relative to when profile views are presented.

(3) The shunt cause step: We control the size of the pictures so that it cannot happen that front-on views are larger than profile views, possibly causing a fear response, making *P* true in the possible event that *H* is false.

(4) The experiment step: With the guidance of a statistician, we determine how many research subjects to randomly select for the experiment. To make systematic errors negligible, we employ psychologist-coders who are trained and experienced in administering the FACS to measure facial responses. And we don't tell the research subjects the purpose of the experiment. That could bias their reactions to the pictures, putting the idea in them that we expect the research subjects to show more fear to front-on views than to side views. Last, we perform the experiment, getting the fact, *F*.

(5) The decision step: In describing what we can justifiably conclude, keep in mind the following:

Deduction of *P* from *H*: We believe it is true.

Shunt causes: We controlled all we could think of.

Systematic error: We minimized the size of the systematic error in the experiment.

Sample size: We told the collaborating statistician the three specifications we want to meet concerning the statistical hypothesis test we will perform: (1) the significance level, (2) the effect size, and (3) the power of the test. This means that should the experiment show a difference between the predicted fact, *P*, and the actual fact, *F*, likely it is a real difference, not a statistical fluke. And should the experiment show no difference, likely there is no real difference.

So, what can we conclude? That depends on which of the three possible ways the experiment turns out:

Possibility 1. By a statistically significant degree, on average the pictures of front-on views produce a greater facial fear response than do the pictures of profile views. Hence, the test prediction is true, i.e., $P = T$. Our conclusion would be along the lines of: "Our research supports the hypothesis that people, by and large, are innately more comfortable in seeing profile views of dangerous wild animals than seeing front-on views."

Possibility 2. There is no statistically significant difference in the average responses to seeing front-on views and profile views. The test prediction is false, i.e., $P = F$, and we conclude that the hypothesis is false. This assumes that the power of the statistical test is high, so that the sample size was large enough to have a good chance of detecting a difference if one actually exists. Our conclusion would be along the lines of: "Our research suggests that people are innately equally comfortable in seeing profile views and front-on views of dangerous wild animals." This outcome would set us to reexamining why, in the first place, we thought people would be more comfortable with profile views.

Possibility 3. By a statistically significant degree, on average the pictures of profile views produce a greater facial fear response than do the pictures of front-on views. The test prediction is false, i.e., $P = F$. Our conclusion would be along the lines of: "Our research disproves the hypothesis that people are innately more comfortable in seeing profile views of dangerous wild animals than in seeing front-on views. Morever, it unexpectedly suggests that the opposite may be true, that people are innately more comfortable in seeing front-on views." This queer outcome is the kind that would spur theorizing and lead to conceiving novel hypotheses to explain it.

Comment: When a statistical test fails to reject the null hypothesis of no difference, the view given in some books on statistical inference is that researchers should not claim that there really is no difference overall, beyond the sample data. However, we made the power of the test (and hence the sample size) large. So the likelihood is that in general, beyond the sample data, there is no real difference to speak of between responses to seeing front-on views and profile views.

When a hypothesis cannot be tested with the H-D method. Almost all hypotheses can be tested using the H-D method. What prevents those that can't be?

(1) One thing that can prevent a H-D test is the impossibility of deducing a test prediction from the hypothesis. This is why we will never see a H-D test of the hypothesis, "God exists." The God hypothesis is in the same category as the phlogiston hypothesis. Neither has any deductive connection to P-plane facts. The answer to "Does God exist?" cannot be found through science.

(2) Another thing that can prevent a H-D test, even when it is possible to deduce a test prediction from the hypothesis, is that researchers are unable to figure out how to do it. To date, cosmologists have been unable to test the hypothesis known as string theory, because they have not seen how to deduce testable predictions from it. That is not to say that they will never see how. Some day they may.

(3) Another thing that can prevent a H-D test is whenever a test prediction has been deduced from the hypothesis, but the experiment to get the fact to compare the prediction to is too costly to perform. Physicists and astronomers frequently run up against this. I'll tell you one instance. The hypothesis that gravity waves exist does not lack test predictions, but rather has lacked money to do the experiment.

How making a faulty "If *H*, then *P*" deduction invalidates a H-D test. In more detail than given earlier in this chapter, when we discussed H-D testing with everyday examples, let's look again at how a faulty deduction of *P* from *H*, making the "If *H* then *P*" idea false, makes the H-D test invalid. There's an instance from medieval times when certain religious leaders wanted to test the following hypothesis:

H: Earth rotates on its axis.

From *H* they deduced (erroneously) the following test prediction:

P: There will be great winds from the east and the birds will be swept away.

That is, the "If *H*, then *P*" idea is "If Earth rotates on its axis, then there will be great winds from the east and the birds will be swept away."

Everybody at the time saw the fact that there aren't great winds from the east and the birds aren't swept away. Believing that the test prediction was verified in fact, they declared *H* false, and the H-D test was taken as legitimate support for the idea that Earth does not rotate. Heilbron (1999) writes: "Common sense observed that the earth's spinning, which Copernicus proposed as an explanation of the alternation of day and night and the motion of the stars, would cause great easterly winds and leave the birds behind. Theologians objected that Scripture expressly stated that the sun moves, for otherwise Joshua would not have commanded it to stand still."

The medievals' problem was that the "If *H*, then *P*" idea is false, instead of true as it must be for valid H-D testing. It is just that at the time of Copernicus it was thought to be true. The air surrounding Earth was believed to extend throughout the heavens. Conditioned on believing this, the if-then idea that "If Earth rotates on its axis, then there will be great winds from the east and the birds will be swept away" was incorrectly accepted as true.

Don't assume that the same cannot happen today. Just conceivably, some of yesterday's knowledge may come to be viewed as mistaken by the knowledge of today. And with that, deductions once based on the mistaken knowledge and believed to be timelessly true may be seen as false.

4.3 A list of research examples illustrating the H-D method

The following illustrations of how researchers have applied the H-D method should give you ideas for applying it in your research. For the sake of brevity, we will not mention every consideration of the applications, such as steps the researchers took to keep shunt causes from acting, took to set the required sample sizes, or took to limit systematic error. Yet read the cited reports behind the illustrations and you will see that best research practices were followed.

Res. Ex. 4.1: How was the Doppler effect discovered?

After casually listening to noises made by moving trolleys and trains in Vienna, Christian Doppler proposed this hypothesis:

H: When a source of sound moves toward the observer, the sound wave is compressed into a shorter wavelength, and when the source moves away from the observer, the sound wave is stretched into a longer wavelength.

The test prediction he devised relies upon the pitch of sound as a measure of the degree of compression or stretching. His plan: Have a group of musicians stand on a train station platform and play a note of a given pitch. Have a second group riding on a train moving toward and then past the platform play the same pitch. Standing beside the platform, have a third group of musicians, selected for their perfect sense of pitch, listen for dissonance, clashing pitches. If *H* is true, he reasoned, the following test prediction, *P*, must also be true:

P: A dissonance will be heard as the train approaches the platform, and a different dissonance will be heard as the train moves away.

Doppler did the experiment. The test prediction was true, corroborating the hypothesis.

In the 1840s when Doppler did this research, wavelengths of sound were unmeasurable constructs, i.e., isolates. An indirect test of the hypothesis was necessary because microphone pickups hooked to oscilloscopes, which can measure sound waves directly, had not yet been invented.

Res. Ex. 4.2: Is our hearing immediately adaptable?

Our eyes are part of our visual system, and our brains are the rest. Equip chickens or people with prism glasses, altering visual locations on their retinas, and in weeks their brains adapt. The chickens are accurately able to locate grains of food; the people are accurately able to locate small objects. Remove the prism glasses and immediately there is fumbling. Only gradually do chickens and people adapt back.

With the human auditory system, our ears are one part of it and our brains are the rest. Hofman et al. (1998) decided to test the following hypothesis, suggested by analogy with the visual system:

H: People who learn to locate sources of sounds when the normal sound cues have been altered will, when the normal sound cues are restored, only gradually adapt back.

The test prediction involved fitting four research subjects with molded ears, which alter the locations of normal sound cues, and having the subjects wear the ears in a laboratory setting until they could correctly locate the sources of sounds from audio speakers arranged about them. Specifically, the test prediction is this:

P: Immediately after the molded ears are removed, the subjects will misidentify the locations of sources of sounds. Only gradually will their previous ability to correctly identify the locations return.

Hofman et al. performed the experiment. The fact was that as soon as the molded ears were removed the subjects accurately located the sources of sounds. Contrary to expectations, the test prediction was false, making the hypothesis false. An editorial commentary accompanying Hofman et al.'s article notes that "It's as if the listeners had learned a new 'language' and now had two sets of ears with which they were proficient."

Concerning the research on the auditory system, I want to mention two points:

(1) How did Hofman et al. conceive the research hypothesis? Was it by the retroductive method? The question method? The subject method? Answer: The question method. They asked themselves whether a truth about the visual system was true of the auditory system. Their tentative answer, i.e., their hypothesis, was that it was, and so they tested their hypothesis with the H-D method.

(2) The research hypothesis and the test prediction look somewhat the same. But they are far from it. The hypothesis is general. It holds for sources of sound of any type, anywhere. The test prediction is about a specific source of sound, in a specific setting. The hypothesis, because of its generality, cannot be subjected to direct measurement. It is the test prediction, deduced from the hypothesis, that can be subjected to direct measurement. The discovery that it was false to experimental fact means that the general idea – the hypothesis – about any source of sound, anyplace, is false.

Res. Ex. 4.3: How did Vincent Dethier research the biology of the fly?

With a series of H-D tests, Vincent Dethier researched the biology of the fly. He describes them for lay people in *To Know a Fly* (Dethier 1962). The first is a H-D test he designed to answer the question, "How do flies taste their food?". His hypothesis was this:

H: Flies taste their food with their feet.

Evolutionary theory supports this hypothesis. Eons ago, flies that rolled the two acts into one, (1) landing on apparent food and (2) tasting it with their feet, knowing at once whether or not to eat, would have had an edge in efficiency over flies that performed the acts separately, landing and then tasting with their proboscis. Efficiency from doing the two at once would have freed up time for searching for sources of food. And by minimizing the time spent on the food, efficiency would have possibly made it less likely the flies would, in certain instances, lose their lives to predators.

On the assumption that the hypothesis is true, Dethier deduced a test prediction that would have to be true, namely:

P: Flies, attached to a wire with a dab of wax, and gently lowered to the surface of a sugar-water solution in a beaker, will extend their proboscis the moment their feet touch the surface, and start eating.

To control a possible shunt cause, Dethier used non-thirsty flies. Moments before each experimental trial, he allowed the fly to drink all the water it wanted. That way, when its feet touched the surface, should its proboscis go down, he would know the fly was not after the water but after the food, the sugar in the water.

Fig. 4.3 is a sketch of the hypothesis, the test prediction, and the two possible experimental outcomes, facts F_1 and F_2. Dethier replicated the experiment with a number of flies. Invariably the actual fact was F_1. When a fly's feet touched the sugar water, it flicked its proboscis to the surface and began eating. Thus the prediction was true, and Dethier considered the hypothesis to be strongly corroborated. Before the H-D testing, it was a bit speculative that flies taste their food with their feet. After the H-D testing, there was little doubt about it.

H: Flies taste their food with their feet.

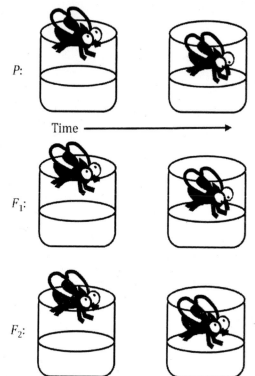

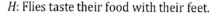

Figure 4.3. A sketch of a H-D test performed by Dethier (1962). He found the experimental fact was F_1. The prediction P agreed with F_1, which corroborated H. Had instead the experimental fact been F_2, P would have disagreed with it, making P false; thus H would have been false.

To Know a Fly describes 13 H-D tests, linked in sequence. The test outcome of one led Dethier to the next question, and from there to the next hypothesis to test. It is a good exercise to read the book and, as you go, note each hypothesis, write down its corresponding test prediction, any control of shunt causes, the fact of the experiment, and whether the hypothesis was corroborated or disproved. Here are the 13 hypotheses in the book (including the one above), each stated after the "if" of an if-then deduction, waiting for you to fill in the test prediction, P, that Dethier used, or to create your own:

1. If (H) flies taste their food with their feet, then (P) . . .
2. If (H) flies' sense of taste is more acute than humans', then (P) . . .
3. If (H) flies dislike salt, then (P) . . .
4. If (H) only the tips of the hairs on flies' feet can sense taste, then (P) . . .
5. If (H) flies have different nerve cells for sensing water, bending, and sugar, then (P). . .
6. If (H) there is a causal relationship between flies' blood sugar level and their hunger, then (P) . . .

7. If (*H*) there is a causal relationship between food in the gut of flies and their hunger, then (*P*) . . .

8. If (*H*) there is a causal relationship between flies' nervous system in the region of the neck and their hunger, then (*P*) . . .

9. If (*H*) salt stimulates flies' thirst through osmotic pressure, then (*P*) . . .

10. If (*H*) flies' thirst is decreased by increased body fluid pressure, then (*P*) . . .

11. If (*H*) female flies need protein to develop their eggs, then (*P*) . . .

12. If (*H*) pregnant flies' preference for protein is caused by egg development, then (*P*) . . .

13. If (*H*) pregnant flies' preference for protein is caused by hormonal control, then (*P*) . . .

Vincent Dethier believed that scientists, besides conceiving theories, should test their theories. "Theories are fine tools to understanding," he said in *To Know a Fly*, "but are not in themselves contributions to truth. Any clever scientist can sit down, marshal the facts at hand, and bounce out of his armchair with a theory [i.e., a research hypothesis]. The scientist who is great is the one who proposes a theory and then attempts to prove or disprove it rather than the one who proposes a theory and then goes off grinning to greener pastures leaving the onerous job of proof or disproof to others."

Vincent Dethier practiced conceiving the research problem that would, upon solving it, point to the next related problem to attack. Thus his career, instead of producing isolated pieces of knowledge about the fly, produced related knowledge, like a completed jigsaw puzzle. Another with the same talent was the physicist J. J. Thompson. About him, Weinberg (2003) wrote that Thompson "was actually not skillful in the execution of experiments. . . . His talent – one that is for both theorists and experimentalists the most important – lay instead in knowing at every moment what was the next problem to be attacked."

Bypassing what is not yet understood. Sometimes scientists have to bypass what is not understood, or they would be stuck there forever. Hear Weinberg (2003): "It is ironic that we still do not have a detailed understanding of frictional electrification, even though it was the first of all electrical phenomena to be studied scientifically. But that is often the way science progresses – not by solving every problem presented by nature, but by selecting problems that are as free as possible from irrelevant complications and that therefore provide opportunities to get at the fundamental principles that underlie physical phenomena."

Res. Ex. 4.4: Do ants have a built-in pedometer?

This one doubly astonishes me to no end. It astonishes me in the knowledge that is discovered, viz., that ants can count into the high hundreds. And it astonishes me in the clever H-D test that produced the knowledge. When you see how it was done, the natural thought is: "Why didn't I think of that?"

To begin with, ants go out from their nest in search of sources of food. When they find one, how do they know how far to travel back to be at the vicinity of the nest, where they can locally explore for its exact location? As long as 100 years ago, a hypothesis that was suggested is this:

H: As ants are going out from their nest, they are counting their steps to where they find food; in traveling back, they count their steps, stopping when the count reaches the number of steps they took going out (as to what direction to follow traveling back, it is known they get it from their celestial compass).

To learn directly whether this hypothesis is true or not, we would have to teach the ants how to talk, and then induce them to tell us how they do it. That is the beauty of the H-D method; its indirect way of learning avoids the impossible.

For the H-D test, Wittlinger et al. (2006) conceived this test prediction:

P: Ants fitted at the food source with stilts made of pig bristles, elongating their legs, will overshoot the location of their nest when they return to it, and ants with legs shortened by clipping will undershoot the location.

For the experiment to get the facts to find out whether the prediction was true or false, Wittlinger et al. trained a group of ants to walk in a straight metal channel, ten meters long, which connected their nest with a feeding site. After the ants had picked up food, the researchers put them in a longer parallel channel, where the ants turned around and made a direct return with the food to their nests. When the ants were accustomed to this circuit, the researchers intervened by catching the ants that reached the feeder. Some got stilts, increasing their stride length; some had their legs shortened, decreasing their stride length. When given a crumb of food and released to return home to their nest, 10 meters away, those with stilts traveled an average of 15 meters back before looking for the nest. Those with shorten legs traveled 6 meters back before looking for the nest.

Thus the test prediction was true, corroborating the hypothesis that ants count their steps. So, this is very probably how it is, and any other hypothesis is very probably not how it is, such as ants keeping track of the energy they burn in walking, or ants memorizing the locations of certain features they pass in going out for food, and then taking the features as guideposts in returning.

Res. Ex. 4.5: Is grammar "hard wired" in our brains?

This one is a research treasure. Coppola and Newport (2005) conceived a test of Noam Chomsky's theory that grammar is innate, the result of the architecture of our brains, not learned. They did this by conceiving a test of a specific hypothesis that is a consequence of Chomsky's theory, namely:

H: The grammatical category of Subject is innate.

This means that people, in beginning to tell a story, will innately not keep listeners in the dark about the story's subject. They will not typically begin like this, "It's valued. It's deserved by all. It's worth fighting for." – not mentioning until the end what "it" is, e.g., liberty.

To test *H*, Coppola and Newport assumed that it is true, and on this condition deduced from it the following test prediction, *P*:

P: When people who have never been in contact with conventional languages tell their family members of events they have seen, they will begin by giving the subject.

The experiment to obtain the fact, *F*, involved three deaf Nicaraguans. Hear Coppola and Newport: "These deaf individuals have had no contact with any conventional language, spoken or signed. To communicate with their families, they have each developed a gestural communication system within the home called 'home sign.' Our analysis focused on whether these systems show evidence of the grammatical category of Subject."

Accordingly, the three were shown 66 videotaped events, and afterwards in "home sign" language they told their family members about the events. The three always started with the subject. From these experiments, *P* was true. That corroborates *H*. In turn, that lends support to Chomsky's theory.

Res. Ex. 4.6: How do Clark's nutcrackers find their cached seeds?

A Clark's nutcracker collects piñon-pine seeds and buries them in thousands of places. Throughout the winter, it comes back to the buried caches, digs them up, and eats. How does it find the caches? Smell? Random searching? What? As described by Vander Wall (1982) and Shettleworth (1983), here is a hypothesis, a test prediction deduced from it, and the experimental result:

H: Clark's nutcrackers memorize the locations of their caches.

P: The birds will more quickly find seeds they cached than find seeds the researcher has secretly cached at locations near to their caches.

The experiment showed that *P* was true. So, let it be said loud and clear: It's true that Clark's nutcrackers memorize the locations of their caches.

Res. Ex. 4.7: How does the sand scorpion detect its prey?

To zero in on the answer, Brownell (1984) went through a series of proposing and testing six hypotheses. The result of one H-D test guided the choice of the next hypothesis to test. Here are the six hypotheses with corresponding test predictions:

H-D test 1. *H*: Scorpions rely on vision to detect their prey. *P*: Unsighted scorpions (eyes temporarily covered) will be unable to detect prey. Replicated experiments showed that *P* was false, disproving *H*.

H-D test 2. *H*: Scorpions rely on sound to detect their prey. *P*: With the sound of prey blocked, scorpions will not react to nearby prey. Replicated experiments showed that *P* was false, disproving *H*.

H-D test 3. *H*: Scorpions rely on ground vibrations from the movement of prey to detect their prey. *P*: With ground vibrations blocked, scorpions will not react to nearby prey. Replicated experiments showed that *P* was true, corroborating *H*.

Brownell next directed his attention to understanding how scorpions detect ground vibrations. He hypothesized a plausible mechanism and created a test prediction for it:

H-D test 4. *H*: Scorpions rely on their tarsal slits to sense ground vibrations from prey. *P*: With the slits on all legs made inoperative by puncturing them with a fine pin, scorpions will not react to ground vibrations from nearby prey. Replicated experiments showed that *P* was true, corroborating *H*. This piqued Brownell's curiosity to discover how scorpions estimate the direction to their prey. First, he guessed that the intensity of ground waves is critical:

H-D test 5. *H*: Scorpions estimate the direction to prey by detecting the differential intensity of ground wave vibrations across their feet. *P*: When the intensity of ground waves is varied, scorpions will misjudge the direction to their prey. Replicated experiments showed that *P* was false, disproving *H*. Brownell had guessed wrong. He guessed again:

H-D test 6. *H*: Scorpions estimate the direction to prey by detecting the time delay of ground wave vibrations across their feet. *P*: When the arrival times of ground waves at the scorpions' feet are manipulated, they will misjudge the direction to their prey. Replicated experiments showed *P* was true, corroborating *H*.

And that's that. With these six H-D tests, knowledge about scorpions was increased in three respects of catching prey. (1) Scorpions sense the presence of prey by sensing ground vibrations set off by prey. (2) Scorpions detect the ground vibrations with their tarsal slits. (3) Scorpions locate the direction to prey by detecting the time delay of ground vibrations crossing their feet.

Practicing border-crossing thinking. There are classes of similar research hypotheses. To take one example, the hypothesis of scorpions locating the direction to prey by sensing a gradient of ground waves with their feet is in the same class as the hypothesis of snakes locating the direction to prey by sensing a gradient of chemicals with their forked tongue (described in Res. Ex. 3.31). Moreover, there are classes of similar test predictions. To mention just one, frequently in communications studies with animals, plants, or people as subjects, a test prediction will involve augmenting a signal that is naturally sent from one subject to another, or blocking it entirely.

As good comedians borrow good jokes from each other, so do good scientists borrow good ideas from each other. Why not do this then? As you read reports of research in all fields, record in notebooks the hypotheses you come across, putting them into classes, like hypotheses with like. Do the same with test predictions. Make your notebooks grow with your career, and borrow ideas from them for your research. You will be in company with Charles Darwin. He adapted a theory about the geologic processes that build mountains, to get a theory that accounted for the building of ring-shaped coral reefs in the Pacific and Indian Ocean basin (Browne 1995).

Res. Ex. 4.8: What influences do predators have on a food web?

Borrowing a theory from plant ecology, Power (1990) proposed a hypothesis concerning plant life in a river:

H: In aquatic food webs with odd numbers of trophic levels (3, 5, etc.), the supply of nutrients limits the size of the plant population at the bottom level. And in aquatic food webs with even numbers of trophic levels (2, 4, etc.), the presence of large fish at the top level limits the size of the plant population at the bottom level.

From *H*, Power deduced the following test prediction:

P: In 3-trophic webs of (1) insect larvae feeding on (2) midges, and midges feeding on (3) algae, the algae should be nutrient-limited (specifically, nitrogen-limited). In 4-trophic webs of (1) steelhead trout feeding on (2) insect larvae, the larvae feeding on (3) midges, and the midges on (4) algae, the algae should be limited by the trout.

Power did the experiment on the Eel River in California. For the 4-trophic case, she left a section of the river as it was, fencing it with plastic mesh. For the 3-trophic case, she took a section like the first, but inside the fence of plastic mesh she removed the trout. From the experiment she found that *P* was true, which corroborated *H*.

Of special note: Initially *H* had some credibility because it was borrowed from a credible theory of plant ecology, and the H-D test raised its credibility further.

Res. Ex. 4.9: What informs kleptoparasites that it is a safe time to steal?

Certain spiders, called kleptoparasites, venture onto a host spider's web to steal food the host has wrapped and hung for later eating. To reduce the risk of becoming a meal itself, the kleptoparasite stays off the web when it senses there is no food there to steal. It rigs signal lines running from its resting place, 20 to 30 cm outside the web, to the web. With these lines, it reads the vibratory pattern of the host's prey-catching sequence, and it adjusts its stealing behavior accordingly. Which event(s) of the host, among the chain of events shown in Fig. 4.4, informs the kleptoparasite that a trip to the web's hub will probably be safe?

To answer this, Vollrath (1979) designed a series of H-D tests. The first test promised to eliminate about half of the chain of events. For the part of the chain that remained, he designed a second H-D test, eliminating about half of the remaining chain. With another H-D test, he had located the single event that informs the kleptoparasite of when it will soon be safe to venture onto the web. It is the vibrations from the host spider wrapping its prey.

The most reliable methods of knowing – collectively called the methods of science – are amazing. As I like to say, we cannot teach kleptoparasites to talk and tell us how they know when it is safe to venture onto spider webs. No matter, we can use the H-D method to know.

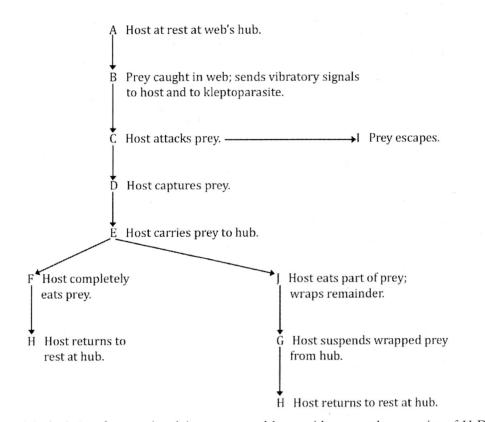

Figure 4.4. A chain of events involving a prey and host spider, central to a series of H-D tests done by Vollrath (1979).

Vollrath's research on kleptoparasites illustrates border-crossing thinking. He borrowed a technique that is common to computer science but uncommon to ecology. In computer science, the technique is called the split-half search method. It is an efficient way to locate an item in a list. For instance, to identify a word that someone has secretly marked in a book, ask whether or not the word is in the first half of the book. Learning which half it is in, ask whether or not it is in the first half of that half. Keep going, and for a 2,000 page dictionary, 11 questions will locate the page. Continue on that page, asking whether the word is in the left or right column. With several more questions you will have located the sentence containing the word. Ask which half of the sentence it is in, and so on, splitting the remainder in half and searching one of the halves.

Res. Ex. 4.10: Can climate affect Earth's geologic behavior?

Hardarson and Fitton (1991) tested a mechanistic theory about cause and effect which may be stated this way:

H: The warming climate that ended the last ice age melted the 2-kilometer-thick ice cap covering Iceland, which lowered the pressure within Earth's mantle, which caused rock in the mantle to melt.

What if this is true? Then the geologic record would hold indications of cause and effect. Specifically, this prediction would be true:

P: Concentrations of certain elements in basaltic rock samples that formed when the ice age ended will indicate that melting of the mantle increased.

Hardarson and Fitton determined the fact. Chemical analysis of rock samples indicated that the amount of melted material in the mantle doubled as the ice age was ending. The prediction was matched by the fact, corroborating the hypothesis.

4.4 The benefits of performing multiple H-D tests of a hypothesis

"Multiple" refers to two or more H-D tests. We deduce from a hypothesis, H, two, three, or more test predictions, diverse in their natures. Then we perform separate experiments to get the facts to judge each test prediction against. A student of mine once offered this illustration of the approach:

H: Rodents nesting under *Larrea tridentata* destroy seedlings of invading plant shrubs.
From H, the student deduced three diverse test predictions:

P_1: Relatively few shrubs will be found growing beneath *Larrea* having rodent burrows underneath them.

P_2: *Larrea* without rodent burrows underneath will have more seedlings growing beneath them than will *Larrea* with burrows.

P_3: Seedlings inside cages to protect them from rodents will survive as well underneath *Larrea* as in places away from *Larrea*.

The three test predictions are diverse. The first is about shrubs. The second is about rodent burrows, or not, and unprotected seedlings. The third is about seedlings protected from rodents eating them. The next step of the plan is to perform three separate experiments. The result will be one of three outcomes:

Outcome 1. All of the P_i are true. This strengthens belief that H is true, relative to having fewer test predictions, e.g., any two of the above. Because, for the test predictions all to be true and yet for H actually to be false, unknown shunt causes would have to be making the P_i true in each test. It little stretches the imagination to believe that an unknown shunt cause may be operating in one of the tests. It stretches it more to believe they are operating in two of the tests. It stretches it even more to believe they are operating in all three of the tests. As a rule, the more P_i there are, all of them true, then the more likely it is that an unknown shunt cause is not operating in at least one of the H-D tests. Knowing that will make us quite certain that the hypothesis should be considered true.

Outcome 2. All of the P_i are false. As it takes just one false test prediction to be reasonably sure that the hypothesis is false, having them all false will make us extremely certain that the hypothesis is false.

Outcome 3. Some P_i are true, and some are false. Possibly the hypothesis is false but shunt causes are operating and making some P_i be true. Or possibly the hypothesis is true but something is making one or more of the test predictions false, such as an error(s) in experimental procedure, or a mistake in deducing a test prediction(s).

Another illustration of multiple test predictions is from a book by Zee (1989; p. 112):

H: Some galaxies are made of matter, and some of antimatter.

Assume, for a moment, that H is true. That would make us expect that the following three test predictions are true:

P_1: There will be antinuclei in cosmic rays coming from antimatter galaxies.

P_2: There will be a tremendous lot of energy being produced in the collision of galaxies, where one is made of matter and the other of antimatter.

P_3: Physicists would doubtlessly have discovered a mechanism explaining the segregation into galactic clumps of matter and antimatter.

The facts are contrary to these test predictions, making them all false. Each disproves H, yet each is not an absolute knockout. But three almost-knockouts add up to an absolute knockout, forcing us to discard the hypothesis.

Res. Ex. 4.11: Human handedness: nurture or nature?

The fact is, about 89.5 percent of all people prefer to use their right hand to throw objects. Why this preponderance of right-handers? Retroductively, Calvin (1983) hypothesized that it is innate, the consequence of our genes.

The support Calvin gives runs this way: Genes for right-handedness could have come to dominate because mothers in ancient times favored cradling their babies with their left arm. This put the baby's ear close to its mother's heart, on the left side of the chest, where the heartbeats pacified her baby. A pacified baby was not fussing, and so was less likely to frighten away small game its mother was hunting by throwing stones with her right hand. This increased its mother's and its chance of surviving. By this, genes for right-handedness grew dominant. Notes Calvin:

"While silent hunting skills may be an important survival skill for mothers themselves, the infant is even more exposed to selection than the mother: . . . [the infant's survival] is threatened by even minor diminution in the mother's health or fitness." Among the studies that Calvin cites for the preference of mothers carrying their babies with the left arm is "a survey of madonna-with-child paintings in European art galleries . . . of over 400 such artworks from four cultures . . . showed left-armed infant carrying in 80 percent of the cases." Because right-handedness prevails across cultures, it is likely to be hard-wired in us, not learned.

Another possible evolutionary force for spreading genes for right-handedness involves warfare. Ancient males who held their shields with their left hand over their hearts had a survival advantage. Naturally, dexterity developed in the right hand and arm, as in doing battle with spears.

Thus ends the retroductive phase of investigating whether or not handedness is in our genes. Taking us the rest of the way are three H-D tests. Specifically, Klar (2003) refined Calvin's hypothesis to this:

H: A person's natural preference to be a righthander or a lefthander is determined by a single gene carried by the person, and this gene also determines the direction that the hair on the back of the person's scalp curls – clockwise or counterclockwise.

With the help of a model of genetics, Klar deduced from the hypothesis this test prediction:

P_1: More than 90 percent of right-handed people have scalp hair that curls clockwise. Of left-handed people, half have hair that curls clockwise, and half have hair that curls counterclockwise.

Klar performed an experiment to get the fact. He randomly sampled people, determined whether they were right-handed or left-handed, and determined the direction of their hair curls. The test prediction, P_1, agreed with the experimental fact, consistent with the hypothesis being true.

Nor is this all. Klar deduced two other test predictions, P_2 and P_3, from the hypothesis, performed the experiments, and the test predictions agreed with the facts, twice again corroborating the hypothesis (see his article for the details).

To recap, the first part of this research is retroductive. The fact is that most people are right-handed. A hypothesis was retroduced to explain the fact, namely that our genes are responsible. The hypothesis was supported with reasons involving hunting and warfare. The second part of the research is based on multiple H-D tests of the hypothesis: Three different H-D tests separately corroborated the hypothesis. That cinches it: genetic predisposition causes the majority of people to favor their right arms and hands.

Res. Ex. 4.12: Who is the Elder Lady?

A mummy of unknown identity was found in Amenhotep II's tomb and became generally known as the "Elder Lady." Who was she in real life? Could she have been Queen Tiye? The facts are that (1) Queen Tiye's mummy had never been identified. (2) Queen Tiye lived at a time that could have resulted in her mummy being brought into Amenhotep II's tomb. (3) The Elder Lady's arms were crossed, indicating she was of great importance, and Queen Tiye was of great importance. Consequently, Harris et al. (1978) retroductively proposed this hypothesis:

H: The Elder Lady is the mummy of Queen Tiye.

This completed the retroductive part of the research, generating some belief that the hypothesis is true. The second part involved two diverse H-D tests. Concerning the first, the Egyptian Museum held the mummies of 10 queens whose identities were known. One of these was the mummy of Queen Thuya, the mother of Queen Tiye. The cephalograms (lateral head radio-grams) of the 10 were available, and the Elder Lady's was added, making 11. For the cephalograms, Harris et al. measured attributes of their sizes and shapes. To picture this data, picture a matrix with 11 columns, each standing for a cephalogram, with a number of rows in the matrix, each indicating an attribute measured on the cephalograms. Each column of the matrix is a numerical profile of the given mummy's skull shape. As explained below, these raw data became a derived fact for testing the following test prediction, deduced from *H*:

P_1: Of the similarities of cephalograms for the 55 possible pairings of mummies ((11 x 10)/ 2 = 55), that of the Elder Lady and Queen Thuya should be most alike.

Harris et al., performing a cluster analysis, transformed the raw data into a derived fact, viz., a diagram displaying the similarity between the mummies of each pairing (such diagrams are called dendrograms). Closest in similarity were the cephalograms of the Elder Lady and Queen Thuya. This agreed with the prediction, P_1, corroborating *H*. Still, Harris et al. thought that the truth of *H* was too much open to doubt. For one thing, the deduction "If *H*, then P_1" could possibly be mistaken. Based as it is on the probabilities of genetics, it would tend to be true on average, but in any one instance might be in error; daughters do not invariably inherit the morphologic characteristics of their mothers. Secondly, it would be better to have a larger number of mummies to reduce the chance of accidental agreement.

The findings were sufficiently convincing to sway the Egyptian Department of Antiquities to permit a further H-D test of *H*, based on precious hair samples in their care. For the test, Harris et al. deduced a further prediction from *H*:

P_2: A sample of the Elder Lady's hair should match a sample of hair taken from a wood sarcophagus found in King Tutankhamon's tomb and bearing the name of Queen Tiye.

Chemical analysis showed a match. *H* was confirmed. Say no more. Three separate lines of evidence, attaching to the P-plane through unrelated constructs, eliminate practically all doubt in *H*'s truth: (1) The retroduction of the hypothesis from the "arm-crossing" evidence and other facts, (2) the H-D test with a cluster analysis of the cephalograms, and (3) the H-D test with the hair samples. See Romesburg (2004, p. 48-51) for details of the cluster analysis.

Testing to answer "Is nature or nurture the cause?"

Suppose a psychological trait X is observed in a great many people of a culture. Is trait X inborn, or is it culturally learned? A H-D approach to solving the question is to hypothesize that the trait X is inborn. For if the hypothesis is true, the prediction must be true that the trait is present in people of every culture around the world. It remains then to get the fact by searching every culture. Is trait X found in, say, at least 10 percent of every culture's living people? Is it manifest in at least 10 percent of its literature, art, laws, and traditions? (We can't reasonably insist that trait X be anywhere close to

100 percent evident. For even though it may be genetically determined, it may not visibly show in all people's actions and words.)

Follow me on this, where trait X is love of nature. Is love of nature innate or learned? For talking purposes, let's assume that a search discovers that a substantial number of Americans, Brazilians, Egyptians, Indians, Finns, Eskimoes, Africans – all cultures and races – love the forests, the sky, the savannas, the mountains, the deserts, the flowers, the dolphins, the elephants, the turtles, all the animals and plants, indeed all of nature. And without exception this love is recorded in each culture's literature, art, laws, and traditions. Then we must conclude that the hypothesis that love of nature is genetically determined is true. Research reported in the book, *The Biophilia Hypothesis*, does just this (Kellert and Wilson 1993).

Besides love of nature, other genetically set traits that come to mind are love of art, love of science, and love of religion. To take one, science exists in all cultures. Yes, some choose not to develop it in the all-out manner that certain countries do, and do not try to hook their people on science's intrigues. Still, at least crude science, such as developing rules of thumb for predicting weather, has existed everywhere for as long as we know. The evidence, as a whole, supports the hypothesis of an innate constitutional need of humans to discover laws of the natural world.

4.5 The benefits of having control groups in H-D tests

Two separate purposes call for control groups:

One is to investigate cause and effect. There, control groups serve a well-known purpose. The researcher randomly selects a set of experimental units from a population of like experiment units. When they are people or animals, they are called research subjects, or just subjects. Otherwise, they are called experimental units. In any case, the researcher gives half or so of the set a treatment; they are the treatment group. The researcher leaves the rest as they are; they are the control group. The level of response in the control group sets a baseline for judging the level of response in the treatment group. This is the well-known purpose of control groups in cause-effect research (the subject of chapter 5), which does not concern us in this chapter.

In testing research hypotheses with the H-D method, control groups serve a completely different purpose. It's like this: We test a hypothesis and, suppose, the test prediction is true. Is the hypothesis true? We can't be sure because an unknown shunt cause could be making the test prediction be true though the hypothesis is actually false. Those darn unknown shunt causes! If only there was a way to rule them out. Well, there is. Have a control group, provided that is possible. With shunt causes ruled out, a test prediction that is true to an error-free fact proves that the hypothesis it is deduced from is true.

It's too bad it isn't always possible to have a control group. In the H-D test discussed earlier about the dam breaking, there is only the one dam and its river, so no control group is possible. Likewise in the earlier H-D test about genes being responsible for right-handers making up the majority of the population, there can be no gene-free people who could act as a control group. Yet where a control group is possible, don't do a H-D test without it. In certain outcomes of the test you will be able to justifiably say, "Honest to goodness, cross my heart, hope to die – the hypothesis is true." A place to see this is the following research example.

Res. Ex. 4.13: Why do some trees' trunks have elliptical cross-sections?

Put on your forester's garb and come with me into a stand of trees that have grown up with the wind blowing mostly in one direction. Looking around, we see that the horizontal cross-sections of the trees' trunks, at five or so feet above the ground, are approximately elliptical, with the long axis of the ellipse running in the direction of the prevailing wind. By the retroductive method we hypothesize this explanation:

H: Trees raised in a place where the wind blows mostly in a given direction will respond genetically by growing elliptical in cross-section, with the long axis of the ellipse in the direction of the wind.

Support for this hypothesis rests on evolutionary theory. Trees that can best withstand the forces of wind survive in greater numbers than weaker trees. The strongest trees, least prone to breaking, are those that grow elliptical cross-sections with the long axis in the direction of the prevailing winds.

Obviously the truth of H cannot be determined directly. As I like to say, we cannot teach the trees to talk and reveal how they operate. What we can do is test H with the H-D method. (I am repeating myself here, and I know it, but it is necessary.) Suppose we deduce from H this test prediction:

P_1: A group of young maple trees planted in the city cemetery of Logan, Utah, which regularly gets westward winds out of the canyon to its east, will grow up with elliptical cross-sections running east-west.

Let us imagine we plant a group of maples in the cemetery, and return in 30 years and get the facts to see how the prediction fared. We see that the cross-sections are elliptical as predicted. P is true, H is corroborated. Just the same, we would not be comfortably sure that the hypothesis is true. Perhaps an uncontrollable shunt cause – conceivably the east-west path of the sun – somehow made the test prediction be true.

Now suppose that instead of the above H-D test, we do the following one. From H we deduce a test prediction incorporating a control group:

P_2: A group of young maple trees planted in the cemetery, free to bend, will grow up with elliptical cross-sections running east-west. A control group of young maples planted nearby, guyed so they cannot bend, will grow up with round cross-sections.

Should the experiment for this show that P_2 is true, we can be very sure that H is true. By having the control group of guyed trees, we have clamped down on the possibility that one or more known or unknown shunt causes, e.g., the sun moving across the sky, acts to make the test prediction true in the event H is false. (Forest scientists, by the way, have done this experiment at a place other than the cemetery, with a test prediction like P_2, and found it was true.)

What a control group does is clamp down on shunt causes so they cannot operate. Thanks to a control group we can safely bet that trees respond genetically to prevailing winds, by growing elliptical in cross-section, with the long axis of the ellipse in the direction of the wind.

The next research example, illustrating the design of a H-D test of a hypothesis concerning how salmon navigate, puts the spotlight on the benefit of having a control group of salmon.

Figure 4.5. A sketch of a river with branching tributaries entering a lake, for illustrating a H-D test to answer how salmon find their way home to the place X to spawn.

Res. Ex. 4.14: How do salmon find their way home?

Fig. 4.5 depicts a river entering a lake. The river has tributaries, the tributaries have tributaries, and so on, branches off branches. Salmon are born on a tributary at a place marked X. As fingerlings they leave X, swim downstream into the lake, and spend nearly all of their adult lives there. In their final month, they return home to X to spawn. How do they find their way home?

At one time, people thought it might be that as the young salmon swim downstream they memorize key features, and on returning upstream they follow the features as guideposts. For instructive value, let us design a H-D test of that idea:

H: Salmon rely on vision and memory to find their way upstream to X.

There are various test predictions we can deduce from *H*. We will choose one that involves a batch of salmon that are born and marked at X, survive their time in the lake, and are later captured at the river's mouth on their return journey. Half of the captured salmon are fitted with opaque eye caps, and are termed the "blinded" treatment group. The rest are only marked and called the sighted control group. After capture at the river's mouth, both groups are released to continue on upstream, each exposed to the same weather and stream flow conditions. Under the assumption that the hypothesis is true, the following test prediction must be true:

P: The "blinded" treated group of salmon will not make it home to X, while all or nearly all of the sighted control group will.

The reason we allow for not all of the sighted control group getting home is that a few may accidently die on the way.

There are four possible outcomes. The one that occurs will be determined by the experiment when it is run.

Outcome 1. The facts are that the blinded salmon do not get home, and most of the sighted salmon do get home. Hence, *P* = T. This outcome will confirm the hypothesis, i.e., make us very sure that salmon rely on vision to navigate. For, the fact that the sighted salmon get home rules out all known and unknown shunt causes, such as the known shunt causes of a flood or a spill of poison that could keep the blinded from getting home (again, we can't list unknown shunt causes).

Outcome 2. The facts are that all or nearly all of the blinded salmon and the sighted salmon get home. Hence, $P = F$. This means that the hypothesis is false. The salmon must be navigating with a sense other than vision.

Outcome 3. The facts are that neither the blinded salmon nor the sighted salmon get home. Hence $P = F$, which implies that the hypothesis is false. But wait, we can't justifiably claim that it is false. This outcome means that "If H, then P" is not true when tried in this instance. It is false. Something we do not know of is preventing P from following by deduction from H. Perhaps toxic chemicals got into the water, killing all fish.

Outcome 4. The facts are that the blinded salmon get home and the sighted do not. Hence, $P = F$, which means the hypothesis is false. This outcome would make us scratch our heads and send us searching for why. Could it be that procedural errors in doing the experiment caused this seemingly illogical outcome? Could it be that vision is backed up by an extrasensory navigational sense, and the sighted fish got caught by people fishing, while the blinded could not see the lures and continued swimming home, guided by their backup sense? Though far-fetched, unexpected outcomes like this do happen in science and are valuable for galvanizing researchers to search for reasons.

This research example carries a general suggestion. The first thing researchers should do when they start a H-D research project is to examine the conceivable outcomes of the design of their H-D test(s), as this usually stimulates insights for strengthening the design.

4.6 How to test hypotheses economically

There are three sorts of economical H-D tests. (1) Crucial experiments. A crucial experiment tests two hypotheses at once with one experiment, in a way that will support one of the hypotheses and disprove the other. Think of two tests for the price of one experiment. (2) Low-cost experiments. A low-cost experiment tests a hypothesis relatively cheaply. (3) No-cost experiments. A no-cost experiment tests a hypothesis with a fact that is already known, and so is free. Of the three sorts of economical H-D tests, one or more isn't always feasible. But when one is feasible, we should do it.

H-D tests using crucial experiments. I want to describe an everyday crucial experiment, which I pattern on one presented by Carroll (1933). Suppose an elderly diabetic woman, in poor health, lives across the street from you. She's asked you to look in on her twice a day, and on one of your visits you find her slumped over in her chair. Your first hypothesis is, "She's dead." Your second is, "She's asleep or has passed out." If the first is true, you predict that mist won't form when you hold a mirror under her nose. If the second is true, you predict that mist will form. You run the experiment: You hold a small mirror under her nose. Either outcome, seeing mist or not seeing it, will disprove one of the hypotheses and corroborate the other.

The "crucial" in crucial experiment means "decisive." A crucial experiment decides between two rival hypotheses. One gets corroborated, the other gets disproved. ("Crucial experiment" is an old term. The entry for "crucial" in *The Oxford Dictionary of Word Histories* (Chantrell 2002) states that the Latin *crux, cruc-* means "cross," and that travelers coming to a cross or fork in a road faced a choice. The entry goes on to say: "The physicists Newton (1642-1727) and Boyle (1627-91) took up the metaphor [of the travelers' choice] in *experimentum crucis* 'crucial experiment', a decisive test showing which of several hypotheses is correct [corroborated]." A crucial experiment, then, is a special kind of H-D test that tests two hypotheses with one experiment (Platt 1964). There are three steps to designing a crucial experiment:

1. Devise two alternative hypotheses, H_1 and H_2. They might be alternative explanations retroduced to explain a given fact, or alternative answers to a question. It does not matter whether the hypotheses are nearly the same, with parts in common, or are (like the hypotheses in the mist on the mirror test) mutually exclusive and exhaustive.

2. From H_1 and H_2, deduce respective test predictions, P_1 and P_2, that have the following property: if P_1 is true, P_2 must be false; and if P_1 is false, P_2 must be true. (In the event that H_1 and H_2 are retroduced from a fact, P_1 and P_2 are predictions about some other kind of fact.)

3. Do the experiment. The fact, F, of it will either show that P_1 is true and P_2 is false, corroborating H_1 and disproving H_2, or show that P_1 is false and P_2 is true, disproving H_1 and corroborating H_2.

Platt (1964) describes several crucial experiments, including one proposed by Francis Bacon. It begins with the fact that bodies have weight, and there were two hypotheses explaining why, each with proponents in Bacon's time:

H_1: The weight of a body is due to its inherent nature.

H_2: The weight of a body is due to the attraction of Earth's gravity.

To test the two, Bacon devised a crucial experiment. From H_1, he deduced test prediction P_1, and from H_2 deduced test prediction P_2, such that if P_1 were true then P_2 would be false, and if P_1 were false then P_2 would be true.

The test predictions utilized the knowledge that the pull of Earth's gravity on an object weakens as the distance between the object and Earth is increased:

P_1: A pendulum clock and a spring clock will keep the same time when they are lifted to the top of a tall steeple.

P_2: A pendulum clock will run slower than a spring clock when they are lifted to the top of a tall steeple.

The experiment would show what the fact is. Either P_1 would be true and P_2 false, corroborating H_1 and disproving H_2; or P_1 would be false and P_2 true, disproving H_1 and corroborating H_2. According to Platt, Bacon wrote that if the pendulum clock on the steeple "goes more slowly than it did on account of the diminished virtue of its weights . . . we may take the attraction of the mass of the earth as the cause of weight." (The fly in the ointment was that the experiment, sound in logic, suffered from the unavailability of sufficiently accurate clocks.)

A crucial experiment gives us a bargain for our money. We only have to pay for one experiment, not two. And we know that when the experiment is over that one of the hypotheses will be disproved and can be dismissed, and that one will be supported.

Ideally, the surviving hypothesis is made one of the two contenders in a next crucial experiment. In this way, on go the rounds of crucial experiments, like a tournament to select a champion. "Strong inference" is the name that Platt (1964) gave to this efficient tournament strategy of testing hypotheses. It is, however, easier said than done. Single crucial experiments are not rare, but series of crucial experiments in the sense of "strong inference" are. We go now to research examples featuring crucial experiments.

Res. Ex. 4.15: Why does light entering a prism come out colored?

This one exudes amazement! To begin with, from ancient times the fact was known that light entering a glass prism comes out colored. To explain the fact, one camp of ancients retroduced a hypothesis, H_1, and another camp retroduced a different hypothesis, H_2:

H_1: The colors are in the prism; they rub off on the light passing through, making a spectrum.

H_2: The colors are in the light; the prism splits them apart, making a spectrum.

Up until the appearance of Isaac Newton, no one had thought to test either of the hypotheses. Newton tested them both at once by devising a crucial experiment. The two test predictions (diagramed in Fig. 4.6) that he respectively deduced are these:

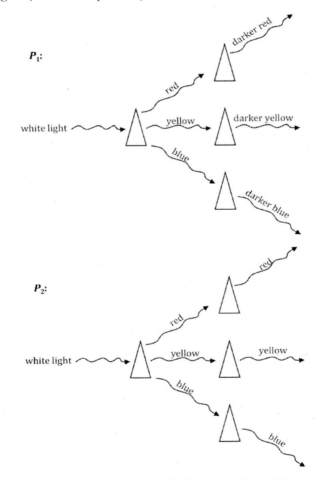

Figure 4.6. A diagram of the two test predictions for Issac Newton's crucial experiment to identify the source of the colors in light that has passed through a glass prism.

P_1: When each of the main colors of light coming out of a prism is passed through another prism, they will emerge darker (since more of the color is rubbed off on the light).

P_2: When each of the main colors coming out of a prism is passed through another prism, each will emerge looking as it did when it entered (since once split, no more splitting is possible).

To do the experiment, Newton (as the story goes) put prisms on the floor of a room so that light out of one would enter another. After darkening the room, he let a beam of light in though a crack between the window frame and the window blind, and it entered the leading prism. The facts he saw with his eyes showed $P_1 = F$, $P_2 = T$. This disproved H_1 and supported H_2.

The design of this crucial experiment is apparent after it is explained. The amazing thing is that it could have been performed much earlier, as prisms had been available since the 11th century. Newton, attacking the problem imaginatively, was the difference.

Res. Ex. 4.16: What did Salvador Luria's Nobel Prize depend on?

It depended on his attending a dance. While watching a man there playing a slot machine, he was struck by the variable size of the payoffs: much of the time no or only a few coins, and a little of the time handfuls. This gave him the idea of how to design a crucial experiment to test two competing hypotheses concerning the phage virus that preys on bacteria (Luria 1984). Underlying the hypotheses were different mechanisms that could conceivably account for the fact that some bacteria prove resistant to phage:

H_1: Resistant bacteria are produced by chance, spontaneous mutations during bacterial growth.

H_2: Resistant bacteria became resistant by contact with phage.

Supposing H_1 was true, Luria deduced that the following test prediction must be true:

P_1: In a set of identical bacterial cultures, the number of resistant bacteria that develop will be quite variable from culture to culture – low in most but spiking high in a few.

In turn, supposing H_2 was true, Luria deduced that the following test prediction must be true:

P_2: In a set of identical bacterial cultures, the number of resistant bacteria that develop will be about the same.

Then it was into the laboratory to do the experiment. Luria prepared a number of cultures and waited for the fact. It happened that P_1 was true, and P_2 was false. H_1 was corroborated and H_2 was falsified. Belief in the mechanism of resistance from contact with phage was out, belief in spontaneous chance mutation was in. This, in turn, spurred a colleague to devise mathematical methods for estimating the spontaneous mutation rates.

Res. Ex. 4.17: Can we control our unconscious mind?

In the course of psychoanalytic research, Weiss (1990) performed a crucial experiment involving patients with repressed unconscious impulses. The two hypotheses:

H_1: People lack all control over their unconscious mental functioning.

H_2: People can exert some control over their unconscious mental functioning.

For each hypothesis, Weiss deduced a test prediction involving how patients would react if the hypothesis was true. For brevity I will omit stating them and just say that the two test predictions meet the condition of a crucial experiment that one of them must be true, and the other must be false. For the experiment, patients received therapy to remove repressions. As a result, this gave them some control over their unconscious reasoning, and certain previously unconscious impulses became vividly conscious to them, without anxiety. Hence, the prediction from H_2 was true, corroborating it.

Res. Ex. 4.18: How do honeybees judge flying distance?

By doing its waggle dance, a foraging bee that has just returned from a source of nectar to its hive communicates the direction and the distance to the source. Yet before Esch et al. (2001) created their crucial experiment, it was a mystery how returning bees knew the distance to waggle about. Two competing hypotheses each seemed reasonable:

H_1: Bees estimate the distance by seeing the features of the landscape pass by as they fly between hive and nectar source.

H_2: Bees estimate the distance by keeping track of the energy they spend in flying between hive and nectar source.

The predictions of each hypothesis involved the following experimental setup. The researchers connected an 8-meter long tunnel (a narrow tube) to the hive. At 3 meters beyond the other end of the tunnel they placed a feeder containing nectar. They painted a pattern on the inside of the tunnel that would give bees flying through it a visually mistaken impression of the actual distance flown, making it appear to them that they flew further than they did. Then, the researchers trained "tunnel bees" to fly from the hive through the tunnel to the feeder. On returning back though the tunnel to the hive, the tunnel bees waggled to their hive mates ("recruit bees"), communicating to them their estimate of the distance to the feeder.

From H_1, the visual hypothesis, the researchers deduced this test prediction:

P_1: When tunnel bees return from the feeder, they will communicate in their waggle dance a longer distance to the feeder than the actual distance to it. Then, recruit bees will fly out this longer distance and search for the feeder there.

From H_2, the energy hypothesis, the researchers deduced this test prediction:

P_2: When tunnel bees return from the feeder, they will communicate in their waggle dance the correct distance to the feeder. Recruit bees will fly out the correct distance.

The facts showed that $P_1 = T$, and $P_2 = F$. The visual hypothesis, H_1, was corroborated; the energy hypothesis, H_2, disproved. The crucial experiment was repeated with the pattern inside the tunnel made more complex to suggest an even longer distance flown. The recruit bees flew way out past the feeder – 72 meters – where they expected to find the feeder, but did not.

Here, the H-D test (in the form of a crucial experiment) does more than corroborate H_1. It makes it thoroughly convincing. For, no shunt causes are imaginable that could make P_1 be true in the event H_1 is really false. And while on this point, for another crucial experiment, well-conceived for clamping down on shunt causes, see Gould's (1985) test of two hypotheses of how honeybees remember the shapes of flowers.

Res. Ex. 4.19: How do green marine turtles choose their nesting beaches?

In the Caribbean Sea and Atlantic Ocean, female green turtles migrate up to several thousand kilometers to beaches on which they lay their eggs. To discover how they choose which beaches to go to, Meylan et al. (1990) designed a crucial experiment to test two hypotheses. H_1 is known as "the natal homing hypothesis," and H_2 as "the social facilitation hypothesis."

H_1: Virgin female green turtles choose to nest at the beach where they were born.

H_2: Virgin female green turtles follow sexually experienced females to a nesting beach, and if the virgins have a successful mating there, in the future they return to that beach to mate.

The respective test predictions, P_1 and P_2, specify different mitochondrial DNA sequences measured in individual turtles living in different colonies. Upon collecting and analyzing the DNA data, the researchers discovered that $P_1 = T$ and $P_2 = F$, which corroborated H_1 and disproved H_2.

Res. Ex. 4.20: How do Pacific brittle stars feed?

Labarbera (1978) designed a crucial experiment to test two hypotheses concerning possible feeding mechanisms by which Pacific brittle stars remove food particles from seawater:

H_1: The Pacific brittle star removes food particles with a sieving mechanism.

H_2: The Pacific brittle star removes food particles with an aerosol filtration mechanism.

The experiment involved feeding brittle stars artificial particles that varied in diameter according to a specified frequency distribution. The test predictions were these:

P_1: The distribution of the particles the brittle star catches will have a characteristic shape set by the limits of a sieving mechanism to remove particles (see the research article for this predicted shape).

P_2: The distribution of the particles the brittle star catches will have a characteristic shape set by the limits of an aerosol filtration mechanism (see the research article for this predicted shape).

The experiment produced the actual distribution. The shape specified by P_1 differed from it, meaning P_1 is false, disproving the sieving idea. At the same time, it tolerably agreed with the shape specified by P_2, meaning P_2 is true, putting the seal of believability on the aerosol filtration idea.

Comment: As the above brittle star research suggests, test predictions need not be about the mean of a frequency distribution. They may be about the shapes of frequency distributions, as well as about properties of frequency distributions other than the mean, e.g., the variance and indices of skewness. For a crucial experiment of two hypotheses of plant competition, where the predictions are about changes in variance and skewness, see Turner and Rabinowitz (1983). Related to this, Gould (1986) noted change over time in the variance of the distribution of major league baseball batting averages, and from this retroductively formed a hypothesis explaining the change.

It is worth remembering that the subject of inferential statistics stemmed, in large part, from agronomists wanting to estimate the mean yields of crops, and today statistics is chiefly sustained by the importance of means in business, management, manufacturing, medicine – indeed, all of the mainly applied sciences. This is why, it seems to me, students in the mainly basic sciences –

biology, ecology, sociology, geology, and the like – are heavily exposed in their statistics courses to testing hypotheses about means. This is good, for testing hypotheses about means is also an essential part of basic science. Yet, students and scientists must keep in mind that in basic science there are opportunities to learn from testing hypotheses whose predictions involve not means but differences in shapes and properties of distributions.

Res. Ex. 4.21: Did the Gestapo perform a mass execution?

A research project of Szibor et al. (1998) in anthropological forensics features a crucial experiment. Central to it is a common grave containing 32 male skeletons, discovered in Magdeburg, Germany, in 1994. Murder was obviously the cause, but who did it was unclear. Historic evidence left the door open to two possibilities:

H_1: The Gestapo performed a mass execution in the spring of 1945, as the war with Germany was ending.

H_2: The Soviet secret police executed Soviet soldiers in the summer of 1953 for refusing to quash a German revolt.

From these hypotheses, the researchers deduced mutually exclusive and exhaustive test predictions. From H_1, they deduced:

P_1: Pollen species in the skulls' nasal cavities will best match those of spring-blooming plants.

From H_2, they deduced:

P_2: Pollen species in the skulls' nasal cavities will best match those of summer-blooming plants.

A pollen analysis performed on the skulls found a high content of pollen species from summer-blooming plants. P_1 was false, disproving H_1; P_2 was true, corroborating H_2. The Soviet secret police, it is a good guess, did the killing.

Res. Ex. 4.22: How did the practice of farming spread across Europe?

Ridley (2000) described the crucial experiment that answered the question. It begins with two rival hypotheses:

H_1: Farmers moved across Europe, taking their farming knowledge with them.

H_2: Farmers stayed put, where they were in Europe, but farming knowledge was copied and by that spread across Europe.

The respective test predictions involved mutually exclusive and exhaustive genetic consequences:

P_1: A genetic gradient exist today across Europe (because races of farmers moved, taking their genes with them, and spreading them in the population).

P_2: No genetic gradient exists across Europe (because races of farmers stayed put, keeping their genes in the places they first settled).

From DNA data on the genetics of people across Europe, a genetic gradient was found, spreading out from the Middle East. So it is that P_2 is false, and H_2 must be dismissed. And because P_1 is true, H_1 is corroborated. The best bet is, therefore, the idea that as the early European population spread out across Europe, farmers moved along too.

Comment: There's a common approach to the above researches about the spread of farming, and about green turtles choosing nesting sites. They rely on genetic information. This brings to mind again the reason for doing border-crossing reading of scientific literature. You may get good ideas you might not get if you confine your horizons to your field.

Res. Ex. 4.23: Was the resurfacing of Venus global or local?

For hundreds of millions of years, meteors have randomly cratered the surface of Venus. At the same time, volcanoes have occasionally erupted, their lava covering craters. How extensive were the eruptions? In answer to this, Hauck et al. (1997) conceived two hypotheses:

H_1: The resurfacing of Venus has been localized, covering craters in local areas but not everywhere.

H_2: The resurfacing has been global, produced by catastrophic volcanic lava flows covering all of Venus.

To test the two hypotheses with a crucial experiment, Hauck et al. deduced the following two test predictions, one of which must be true and the other false:

P_1: Today, the craters will appear as if they have been distributed by a non-random process (because the local wiping out of randomly placed craters, here and there on Venus, makes the craters on all of Venus appear to be distributed by a non-random process).

P_2: Today, the craters will appear as if they have been distributed by a random process (because they were formed by random bombardment since the last catastrophic lava flows).

To test the two hypotheses – local resurfacing versus global resurfacing – Hauck et al. performed computer simulations. They had the computer wipe the imagined surface of Venus clean and then randomly bombard it with craters. Result: one picture of a cratered surface. This they did again and again, 200 times in all. Result: a baseline set of 200 pictures. Comparing properties of the real picture of the surface of Venus with the simulated set of 200 pictures generated by a random process, they concluded that the real picture fit right in with the set. This favors H_2, the hypothesis of global resurfacing.

Res. Ex. 4.24: How did the Witwatersrand Basin's gold form?

From reading the account of the research of Kirk et al. (2002), we learn the following history. Nearly 40 percent of all of the gold ever mined has come out of the Witwatersrand Basin in South Africa. The Basin's sedimentary beds of gold ore have been intensively studied along two lines of constructs: macro-features of the ore, and micro-features. The facts of the macro-feature constructs are that the ore occurs among pebbles marked by the abrasion of windblown sand, with the gold concentrated in the coarsest-grained sediments. To explain these facts, certain researchers retroduced what came to be called the sedimentary placer hypothesis, viz.:

H_1: The gold was formed away from the deposits. Later, the rocks around the gold eroded, and the gold was carried to ancient river deltas in the Basin, underlain by volcanic rock. Still later, volcanic rock flowed over the gold, sandwiching it between the older bottom layer of volcanic rock and the younger top layer.

On the other hand, examination of the gold under a microscope revealed facts of the micro-feature constructs. Namely, the gold crystalized after it was deposited, and the surrounding sediments show signs of hydrothermal activity. To explain this, certain researchers retroduced what came to be called the hydrothermal hypothesis, viz.:

H_2: The gold was formed by hot solutions rising from the sandwich's bottom layer of molten rock as the earth's magma cooled, and later volcanic flows capped it with the top layer.

From the two hypotheses, Kirk et al. deduced the following test predictions:

P_1: (prediction from the sedimentary placer hypothesis): The gold ore will be older than either of the rock layers sandwiching it.

P_2: (prediction from the hydrothermal hypothesis): The gold ore will be the same age as the bottom layer of rock it lies on.

This has the structure of a crucial experiment: one of the test predictions must be true, and one must be false. Yet at the time this crucial experiment was conceived, it could not be performed; laboratory methods for dating the requisite rock samples were too inaccurate. The time came, however, when geochemists could accurately date the bottom and top rock layers. The bottom layer of volcanic rock was dated as 2.89 billion years old, and the top layer as 2.76 billion years old. Then Kirk et al. dated the gold sandwiched between as 3.03 millions years old, older than the bottom and the top rock layers. So, prediction P_1 is true, corroborating H_1, the sedimentary placer hypothesis, and disproving H_2, the hydrothermal hypothesis.

I have found in reading research articles that H-D tests of a single hypothesis far outnumber H-D tests of two hypotheses at once with a crucial experiment. Yet it is a good bet that crucial experiments could be done more. After scientists conceive a hypothesis, they might benefit by continuing to think, trying to conceive a competing hypothesis, and then trying to create test predictions that will allow a crucial experiment.

This finishes our discussion of crucial experiments as a way to test hypotheses economically. We now turn to another approach: constructing low-cost experiments.

H-D tests with low-cost experiments. Let's look at two H-D tests of the same hypothesis, one test with an expensive experiment, one with a low-cost experiment. The hypothesis is that salmon rely upon olfaction to navigate upstream from the mouth of their river, homing in on the characteristic scent of their natal spawning grounds. With an experiment costing tens of thousands of dollars, Scholz et al. (1976) bathed newly hatched salmon in a pond with artificial chemicals, imprinting the chemical's scent in them. Later as fingerlings, the salmon were marked and released to go downstream into Lake Michigan. When it came time for the survivors to return as adults, Scholz et al. released the chemical's scent continuously from a new pond on a tributary that wasn't where their natal pond had been. The prediction was that the salmon would follow the scent upstream and enter the new pond. This proved true. But because so many salmon were lost during the time they lived at Lake Michigan, thousands of fingerlings were required, making it a costly experiment.

Fig. 4.7 shows how an inexpensive H-D test of the olfaction hypothesis was designed, requiring about 150 salmon and a section of stream where two tributaries join (Hasler 1966). The steps of the experiment are these: Capture at site A a batch of adult salmon returning from the sea (or lake), and mark them as A-fish. Capture at site B a batch and mark them as B-fish. Transport both batches to site C. There, randomly select some of each batch to be in a control group, leaving them as they are. The others of each batch become the treatment group; plug their nares so they cannot detect scents in the water. From site C, release the control salmon, letting them swim upstream, and recapture them at site A or site B, wherever each salmon goes to. Do the same with the treated salmon.

If the hypothesis of navigation by olfaction is true, the following is predicted: Most of the control salmon captured at site A will be recaptured at site A, and most of the control salmon captured at site B will be recaptured at site B. Of the treated salmon, a number of those captured at site A will be recaptured at site A and at site B. Likewise, of those captured at site B, a number will be recaptured at site A and at site B.

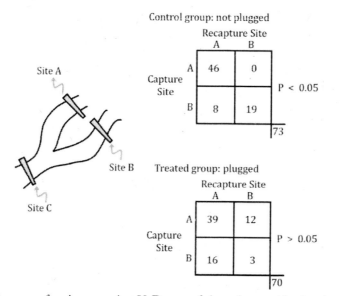

Figure 4.7. A diagram of an inexpensive H-D test of the salmon olfaction hypothesis.

The data from the experiment is in Hasler's (1966; p. 42) book, and is reproduced in Fig. 4.7. For the data on the control salmon (control group), the association between the capture site and the recapture site is statistically significant. For the data on the treated salmon (treated group), the association is not statistically significant. And with that the olfaction hypothesis is quite likely true.

In general, there are various ways to test a hypothesis, and the experiments to do so will differ in cost. A reminder of the obvious then: In the event that all other considerations are about equal, go with the lowest cost experiment.

This finishes our discussion of the approach of constructing low-cost experiments as a way of testing hypotheses economically. There is one more approach to discuss:

H-D tests with already known facts. Concerning the testing of theories with already known facts, listen to the physicist Steven Weinberg (1990): "Very often theories are tested by using them to explain already known facts. After all, when Newton calculated the length of the month in terms of the acceleration of gravity at the earth's surface and the distance of the moon, that [the month's length] was already known, but it was such a good numerical calculation that worked, it was clearly convincing."

Weinberg is commenting on H-D tests where the hypothesis and the test prediction are quantitative. There, the deduction to get the test prediction follows mathematical logic. This prevents us unconsciously fudging the deduction to force a test prediction to match the fact that's already known, obtaining a self-fulfilling prediction.

With a qualitative hypothesis, expressed in words, we run the risk of unconsciously fudging a deduction so that it predicts an already known fact. Words can be subtly twisted, especially under the pressure of wishful thinking to get a test prediction that will agree with the qualitative fact before us.

Res. Ex. 4.25: Paul Dirac's H-D test with an already known fact

In 1928 the physicist Paul Dirac wrote down a handful of mathematical axioms that, he hypothesized, describe aspects of the electron. From them he mathematically deduced a further hypothesis, what now is called the Dirac equation, a wave theory of the electron. The theory predicted that the spin of the electron has a value of one-half, the same value that was known from experiment before Dirac deduced his equation. Physicists everywhere took the match between the prediction and the already known fact as corroborating Dirac's equation.

Let's finish this story about Dirac's equation with some made-up but reasonable degrees of conviction in its truth, expressed as percentages. That is, with the corroboration described above, physicists were 95 percent convinced that Dirac's equation was right. But Dirac noticed that his equation also predicted the existence of anti-particles, and years later they were experimentally discovered, and physicists were then 99 percent sure the equation was right. And when his equation was later joined with another equation and with certain principles of atomic physics, to give a grand equation of quantum electrodynamics having astonishing predictive power, physicists were 99.99999 percent sure it was right. Penrose (1997) writes, "When taken together with the Maxwell electromagnetic equations and the principles of quantum field theory, it [Dirac's equation] provides a scheme of extraordinary power and accuracy – *quantum electrodynamics* – having a precision of some 11 decimal places."

A H-D test of a hypothesis will be valid whenever its test prediction is about an existing fact that the researcher doing the test is ignorant of, but later learns. Likewise, a H-D test will be valid whenever its test prediction is about a fact that is calculated from well known existing data, but no one has the faintest idea what the fact will be until it is calculated. The next two research examples illustrate:

Res. Ex. 4.26: Do wife and husband grow to be alike?

If you stood on a street corner and asked passers-by, probably nine of ten would guess that this hypothesis of cause and effect is true:

H: Shared experiences from long marriages have the effect of making the values and attitudes of wife and husband become more alike.

In place of guesswork, Caspi et al. (1992) applied a reliable way of knowing. To test the hypothesis, they deduced from it this prediction:

P: Examination of an existing database containing the values and attitudes of married couples – economic, aesthetic, political, religious, and so on – assessed with standard psychological questionnaires at the start of their marriages and again 20 years later, will show that couples' values and attitudes are more similar after 20 years than at the beginning of their marriages.

Caspi et al. had no idea what the facts corresponding to the prediction would be when they extracted them from the existing database. To their surprise, they found that the prediction was false, disproving the hypothesis.

(Although the above *H* and *P* may appear nearly the same, there is a world of difference between them. *H* is a general statement. *P* is a particular statement, specifying the kinds of values and attitudes, the length of time, and a selective sample of people.)

Res. Ex. 4.27: Why are there grandmothers?

Gibbons (1997) tells a fact that piqued the interest of evolutionary anthropologists: "Human females are the only ones in the primate family to live well beyond their last pregnancy – often as long as 40 years or more after menopause. . . . Yet, evolutionary theory says that natural selection favors only traits that enhance reproduction – which implies that postreproductive women have no evolutionary reason to live." As a possible explanation of this interesting fact, Hawkes (2004) and Lahdenperä et al (2004) offered this reason:

H: During times long past, women evolved to live to old age to supply food to their grandchildren.

Lahdenperä et al. (2004) devised several H-D tests of this hypothesis. To be brief, I will mention one. The test prediction for it is this:

P: The longer a woman's post-reproductive lifespan (age at death after age 50), the greater the number of grandchildren she leaves in the population.

How does *P* follow by deduction from the hypothesis? Assuming grandmothers living into old age supplied food to their grandchildren, that freed the grandmothers' daughters of some work of gathering food, which kept the daughters' health up, which lengthened their period of fertility, which let them have more babies, which improved the chances of passing on the grandmothers' longer-life genes. (For the considerable details of the support, I refer you to the research of Hawkes (2004) and of Lahdenperä et al (2004)).

To obtain the facts that the test prediction would be compared to, Lahdenperä et al (2004) calculated them from two existing databases of women living in the 18th and 19th centuries.

The facts were that on average the longer a woman's post-reproductive lifespan, the more grandchildren she left in the population.

That the above test prediction and the others not mentioned here were true puts the hypothesis beyond reasonable doubt. The H-D tests were valid because the facts weren't apparent in the existing databases at the time the H-D tests were conceived.

4.7 How progress in science depends on both the retroductive method and the H-D method

The retroductive method (a.k.a. the fact method), the question method, and the subject method are the three methods of conceiving hypotheses. In the following discussion, much of what we will say about the retroductive method applies also to the other two methods. Moreover, when we say "the retroductive method" we will sometimes mean only its application to conceive hypotheses. Other times we will mean its application to conceive hypotheses and then supporting the conceived hypotheses with circumstantial evidence and reason.

With this understood, the message of this section is this: In some areas of discovering knowledge, the retroductive method is superior to the hypothetico-deductive method. In other areas, the hypothetico-deductive method is superior to the retroductive method.

At times in presenting this message, I will mention some ideas of Lipton (2005). He takes the word "accommodation" to mean application of the retroductive method, and the word "prediction" to mean application of the hypothetico-deductive method. As he tells it, "In the case of 'accommodation,' a hypothesis is constructed to fit an observation that has already been made. In the case of 'prediction,' the hypothesis, though it may already be partially based on an existing data set, is formulated before the empirical claim in question is deduced and verified by observation. Well-supported hypotheses often have both accommodations and successful predictions to their credit." (For various comments on the ideas of Lipton (2005), see the Letters section of the 3 June 2005 issue of the journal *Science*.)

Where the retroductive method is superior. Let's first refresh our understanding of the retroductive method. With it, we begin with observing an interesting fact. That stimulates us to conceive a research hypothesis that, in the event it is true, explains the interesting fact. With the same interesting fact, we can apply the retroductive method a number of times, each time conceiving a different research hypothesis, each a possible explanation of the interesting fact.

We now come to the two main ways by which the retroductive method excels in helping science progress:

(1) With the retroductive method, we can in certain cases discover knowledge that cannot be discovered with the H-D method. One instance is the theory of evolution. It was retroductively derived by Darwin, and now is well supported with circumstantial evidence and reason. No matter that it is infeasible to test it thoroughly with the H-D method, it is entirely credible.

Other instances include the hypothesis of why the night sky is dark, the hypothesis of where the Iceman lived and roamed, and the hypothesis of why snakes have forked tongues (as described respectively in Research Examples 3.11, 3.16, and 3.31). Most scientists would swear that these retroduced hypotheses are true. Yet these hypotheses cannot, I surmise, be tested with the hypothetico-deductive method. As a rule, we cannot test some hypotheses because we do not see how to deduce test predictions from them. In others, we can deduce test predictions, but experiments to do the tests are impracticable or impossible.

(2) The retroductive method is instrumental in allowing the H-D method to be applied. The point to bear in mind is that the retroductive method provides hypotheses that become feedstock

for testing with the H-D method. Should the hypotheses then be corroborated with the H-D method, the hypotheses have three sources of credibility: (1) the credibility they have for explaining the fact(s) they are retroduced from, (2) the credibility they have from their circumstantial support, and (3) the credibility they have through their H-D corroboration. To name just one example, Einstein retroductively conceived the theory of special relativity, and gave it circumstantial support. Following this, scientists corroborated it with the H-D method. Moreover, GPS (Global Positioning System) locators are designed to compensate for relativistic effects, and wouldn't work if they didn't. Every time you turn on a GPS locator, you are corroborating the theory of relativity.

Where the hypothetico-deductive method is superior. As I count them, there are ten ways by which the H-D method excels at helping science progress:

(1) When we corroborate a hypothesis with the H-D method, we more strongly believe it is true than when we retroduce a hypothesis. Which would inspire more confidence in a baseball batter's ability – a player who hits a home run over the centerfield wall and then retrospectively points to the wall, or a player who prospectively points to the centerfield wall and then hits a home run over it? And really, which would inspire more confidence in the truth of a hypothesis? Is it when the hypothesis is developed to account for one or more present facts, or when the hypothesis correctly predicts future facts? The H-D method would inspire more confidence.

(2) We can disprove a hypothesis with the H-D method, but we can't with the retroductive method. The H-D method leads to a decision. By it you will decide the hypothesis is either corroborated, or it is false. However, the retroductive method can only lead to the claim that the hypothesis conceived is true. By definition of its steps, it's impossible to retroductively conceive a hypothesis that is believed to be false. There's another way of putting this. The H-D method contains a definite criterion for proving or disproving a hypothesis: it's, "Does it work?" There are two possible answers. One is "yes," one is "no." The retroductive method has no criterion like this. For it, "yes" is its only possible answer to "Does it work?"

(3) When we apply the H-D method, we are testing more than the hypothesis. A test prediction that turns out to be true corroborates more than the hypothesis. It also corroborates the reason(s) that suggested the hypothesis was true in the first place, be that reason a "gut feeling" or a definite theory. Likewise, a test prediction that turns out to be false casts doubt on more than the hypothesis. It calls into question the reason that suggested the hypothesis was true in the first place. In contrast to this, the retroductive method, because it is not a test, cannot cast doubt on ideas that are background to conceiving the hypothesis.

(4) Applying the H-D method is usually more efficient, and hence cheaper, for deciding whether or not a hypothesis is true, than is applying the retroductive method. For just one case of this, recall the research on kleptoparasites that venture onto a host spider's web to steal food the host has wrapped and hung for later eating (Res. Ex. 4.9). Recall that Vollrath (1979) conceived a series of H-D tests to discover the significant event on the web that tells kleptoparasites, "It's now safest for you to venture out to steal food." The H-D tests were based on the half-search method for efficiently locating a significant event in a chain of events. There is no apparent way

of using the half-search method with the retroductive method. Any retroductive approach would be inefficient in locating the significant event. (I can't even imagine how it would go, or if it is possible.)

(5) Generally speaking, the H-D method avoids a wasteful shotgun approach to gathering facts. With the H-D method, we confine the gathering of facts to the kind that bear on the hypothesis. Medawar (1969) speaks to the point: "Our observations no longer range over the universe of observables: they are confined to those that have a bearing on the hypothesis under investigation. . . ."

(6) With the H-D method we choose between different views of how nature possibly is. As Medawar (1969) says: "We carry out experiments more often to discriminate between possibilities than to enlarge the stockpile of information."

(7) With the H-D method, we have more flexibility in applying it than we have with the retroductive method. As Lipton (2005) puts it: "Scientists can often choose their predictions in a way they cannot choose which data to accommodate." Translation: We can choose which test prediction(s), among a number that deductively follow from the hypothesis, to use. Besides this choice, there are other choices Lipton doesn't include. We can choose to clamp down on the known shunt causes, when that is workable. Sometimes we can choose a H-D design that holds promise for clamping down on unknown shunt causes, by having treatment and control groups, and/or multiple test predictions. On these items of control, the retroductive method comes up rather empty handed. Sometimes it doesn't even allow us a choice of the construct(s) we measure, nor allow us to make the measuring process be quite precise and unbiased. Rather, a fact or some phenomena catches our attention, this determines the construct, and we retroduce a hypothesis to explain the fact or phenomena. Overall, then, relative to the retroductive method, the greater choices with the H-D method are more conducive to gaining reliable knowledge.

(8) The H-D method is less susceptible to fudging than is the retroductive method. With the retroductive method, Lipton (2005) takes fudging to be the tendency of scientists to bring too large a domain of facts under the explanatory umbrella of a single hypothesis. But when only some of the facts are relevant, the hypothesis they retroduce is unreliable. They have, in Lipton's word, "fudged" the hypothesis, made it unrealistically sweeping and complex to accommodate the existing facts.

Let me give an illustration of fudging. In the 1980s, a number of parents observed the following two facts. Their child was vaccinated against measles, mumps, and rubella (fact 1) and shortly after autism was diagnosed (fact 2). To explain the two facts, they retroduce the hypothesis that the vaccine can in certain children cause autism. In retrospect, the parents were wrong. Later medical research (Madsen et al. 2003) disproved the hypothesis. The parents had fudged the hypothesis to explain two unrelated facts occurring close together in time – vaccination and autism – as being cause-effect.

For another illustration, there is a card game called Eleusis (invented by Robert Abbott) that impressively demonstrates the tendency to fudge the retroduction of hypotheses. The facts players observe in the game are sequences of cards played that conform in some respect to a secret rule the dealer knows. The players try to retroduce hypotheses that specify the secret rule.

In conceiving a hypothesis based on sequences of cards that conform to the secret rule, players typically fall prey to believing that certain features of the cards and their sequences are part of the secret rule, when actually they are incidental. This leads players to "over hypothesize," i.e., fudge the hypothesis, making it more detailed and complex than the secret rule really is. (For more on this, see Romesburg (1979).)

(9) The H-D method can land us in a new sphere of facts – facts we and our field probably wouldn't have measured for years to come, if ever, had it not been for applying the H-D method. I like to think of this in terms of Trefil's (2005) remark about testing Einstein's theory of relativity, "Relativity says X will happen, so let's measure X." Frequently the measurement, the fact, is one unlikely to ever be measured were it not for the H-D test telling us to measure it. Why? The fact is probably uninteresting. For when the hypothesis says that X will happen, it says so by logical deduction of X from the hypothesis. Yet deduction is unconnected with whether or not X is interesting. There is no such thing as logical deduction that is guaranteed to produce interesting deductions. There is just logical deduction, and what is deduced isn't necessarily interesting.

(10) The H-D method, which is based on deductive reasoning, will take you to ideas that are unreachable by the retroductive method. Deductive reasoning is forward reasoning, where one idea leads forward to the next. Retroductive reasoning (as its name tells us) is backward reasoning, where one idea leads backward to the next. You and I, all of us, are built so that we can reliably reason forward through four, seven, or dozens of steps of deductive logic, not straying into error. Conversely, we are built so that we can only reliably reason backward a few steps. Beyond that we stray into error.

Let me give some details on this. I'll do it in the context of the Big Bang hypothesis of the origin of the universe. Four facts – F_1, F_2, F_3, and F_4 – played the main roles in the story. The first, F_1, was key in retroducing the hypothesis. Once the hypothesis was obtained, the H-D method was applied three times to make predictions of facts F_2, F_3, and F_4. These three facts are so "far out" that, starting with them, it's improbable that anyone could have retroduced the hypothesis, as the chain of backwards reasoning via retroduction would have strayed into undetected error. Let's go through this now carefully.

F_1: All known galaxies are moving away from each other, and the speed at which this is happening is proportional to their distance apart.

From F_1, it was easy thinking to retroduce the Big Bang hypothesis. A child could do it. It was like playing a movie backwards, thinking of the galaxies earlier and earlier in time, until the point was reach when they were all crunched together, the beginning of time.

Once the Big Bang hypothesis was retroduced, it was tested with the H-D method this way: "The Big Bang hypothesis says that hydrogen and helium will be the most abundant elements in the universe in certain predicted proportions, so let's measure their abundances." That was done, giving fact, F_2:

F_2: Hydrogen and helium are indeed the most abundant elements in the universe, with the predicted proportions.

This corroborated the Big Bang hypothesis. Yet it still could be wrong, so it was tested again with the H-D method, this way: "The Big Bang hypothesis says that if we look in all directions into space from Earth, we'll see background radiation made of microwave photons at the same temperature. Therefore, let's measure the microwave background radiation." That was done, giving fact, F_3:

F_3: Looking in all directions into space from Earth, the background radiation is indeed made of microwave photons at the same temperature.

A further test of the Big Bang hypothesis with the H-D method went this way: "The idea of the Big Bang says that if we look we will see stars and galaxies."

The chain of deduction by which the Big Bang hypothesis says this is the following. The Big Bang produced hydrogen and helium, which gravity gathered into clouds, then pulled local clouds together, raising the density to an ignition point, resulting in stars. Gravity then pulled the stars into galaxies. So, the existence of stars and galaxies is a test prediction. And the actual fact, F_4, is:

F_4: Stars and galaxies populate the universe.

Now to the point of this illustration. Logically, it doesn't make any difference whether cosmologists began with one of F_1 or F_2 or F_3 or F_4, and retroduced the Big Bang hypothesis from it, and then used the three remaining facts in three H-D tests of the hypothesis. The end would be the same, that the Big Bang hypothesis is tied down to the P-plane by four different kinds of facts, proving beyond a reasonable doubt that it is true.

But in terms of how easy it is to retroduce the Big Bang hypothesis, it's a breeze to retroduce it from F_1 as it was retroduced, and nearly impossible to retroduce it from F_2 or F_3 or F_4. Starting with F_2 or F_3 or F_4, the chain of backward logic is too long without heading into error. What farfetched thinking: "You know, hydrogen and helium are the most abundant elements in the universe. Aha!, that means there was a Big Bang." Or, what impossible thinking: "Hey, stars and galaxies everywhere. There had to be a Big Bang."

This completes the ten ways by which, to my mind, the H-D method shines in helping science progress. There may be more, and some of the ten may overlap a bit. You be the judge. It's not going to change the conclusion that the H-D method has some unique advantages.

Comment: Opposite to the hypothesis of the Big Bang, the hypothesis of the steady state universe also predicts the existence of stars and galaxies. However, it does not predict microwave photons in all directions in the universe, as the Big Bang hypothesis does. In this difference was the making of a crucial experiment, for it can't be both ways, no photons in all directions, and photons in all directions. When physicists measured the fact, finding photons in all directions, it was "Goodbye steady state, hello Big Bang."

Res. Ex. 4.28: A H-D test of the "snowball Earth hypothesis"

Recall Res. Ex. 3.21 of how the snowball Earth hypothesis was retroduced and supported. Later, Bodiselitsch et al. (2005) tested it with the H-D method. To begin, it is well known that a steady shower of interplanetary dust particles from meteors, asteroids, and comets falls onto Earth today. What's more, scientists believe it was the same a billion years ago and all times from then to now. The particles contain iridium. Some fall on oceans and lakes; some fall on land and wash into oceans and lakes. Either way, incoming iridium ends up in marine sediment that solidifies into rock.

And what if the snowball Earth hypothesis is true? What if there was a period when Earth was iced over? During that time, the interplanetary dust would fall on the ice and get covered with fresh ice. And what if later, as according to the hypothesis, a super-greenhouse environment appeared and, in a geologic blink of an eye, melted the ice? The accumulated interplanetary dust, with its iridium and other elements, would be suddenly released, making its way into marine sediment, then into rock. If this actually happened, there must be a spike in the concentration of iridium preserved in the rock formed at that time, this relative to the concentration during each of the millions of years before and after the presumed worldwide glaciation. Moreover, the longer that Earth was frozen over, the greater would be the concentration of iridium over the years spanned by the spike. And the more suddenly the ice melted, the sharper would be the peak concentration at the tip of the spike.

But what is the fact? Is the predicted spike actual? If it isn't, the hypothesis is wrong. To answer the question, Bodiselitsch et al. analyzed the concentration of iridium along the lengths of three cores of ancient marine sediment, drilled from different places in Africa. The spike predicted from the hypothesis was actually there and revealed "so much iridium deposited at the end of a glaciation 635 million years ago that the planet must have been frozen pretty much solid for 12 million years straight (Kerr 2005)."

This corroboration of the hypothesis is convincing because the researchers were able to rule out a number of shunt causes that could have caused a spike of iridium even if Earth hadn't been frozen over. Bodiselitsch et al. (2005) explain: "Before ascribing an extraterrestrial origin to the Ir [iridium], possible terrestrial sources, including reduced sedimentation rates, increased meteor ablation rates, accretion of extraterrestrial material or of terrestrial dust (such as volcanic airborne particles), and anoxic conditions connected with sulfide precipitation in seawater need to be considered." Continuing, they give reasons that rule out these various shunt causes as possibly causing the spike.

They also argue that the fact of the spike disproves a competing hypothesis called the "slushball Earth hypothesis," which pictures Earth as partially frozen over but free of ice around its equator. Kerr (2005) summarizes the argument: "But a slushball would have melted down within something like a million years as volcanoes belching carbon dioxide fueled a growing greenhouse. 'It's hard to see what would keep a slushball around for 10 or 20 million years,' says climate modeler Raymond Pierrehumbert of the University of Chicago. And even if a slushball did last, its glaciers – unlike those of a snowball – would continually flow down to the sea, steadily depositing iridium, not producing a spike of it." (Although not presented as a crucial experiment, at heart this is one. Two hypotheses – snowball and slushball – each with test predictions such that one must be true and one must be false.)

The retroductive research which led to the snowball Earth hypothesis (described in Res. Ex. 3.21), and the hypothetico-deductive research described here, are independent grounds for believing that Earth was once iced over. As more drill cores are taken in more geographic places, if they all have spikes of iridium, as well as spikes of other elements that can serve as markers, then belief that Earth was iced over will gain more adherents. If it goes the other way, that other drill cores don't have spikes of iridium, and there is no explanation for this that saves the hypothesis, then belief will decline. One way or the other, additional research will bring a consensus about the answer to whether or not Earth was ever completely frozen over.

Comment: In principle, the snowball Earth hypothesis didn't have to be retroduced in the way that is described in Res. Ex. 3.21. Instead, researchers could have drilled cores of ancient marine sediment in different places in Africa, analyzed the concentrations of iridium along the lengths of the cores, seen the fact that there was a spike, and from the fact retroduce the snowball Earth hypothesis. Yet there would be no reason to drill these cores without having the hypothesis first, so researchers wouldn't have drilled them. But even supposing they would have, by looking at the spike, it is very doubtful that they or anyone else would have been able to come up with the snowball Earth hypothesis, for the reason that is point number 10 in the list of 10 points we made above: What preposterous thinking: "Look at these spikes of iridium in these rocks. Eureka!, Earth was once frozen over."

Comment: The value of one piece of evidence can depend on the value of another. It sometimes happens, and I believe it can sometimes be made to happen through how research is designed, that relatively few pieces of evidence – notably a few circumstantial facts and/or reasons supporting a hypothesis, and a corroborating test of the hypothesis – combine to create a smoking gun, leading researchers to believe the hypothesis is true beyond the shadow of a doubt. The few individual items of support, none very strong alone, join to form an interlocking web of support that pins down the truth of the hypothesis. It's like this: Sit on a stool that has one leg and the stool wobbles. Sit on a stool with four or five legs, all close together, essentially touching, and the stool wobbles almost as much as before (a case analogous to various information that is not strictly independent). But sit on a stool with two legs, quite apart, and the stool wobbles only in one direction. And for a stool with three legs, quite apart, in a triangular footprint, the support is firm; and any more legs added do little to make it more firm. The point to bear in mind is that it's not just the amount of supporting evidence that counts; it's the way the individual pieces of it interlock to pin down the hypothesis. Hence, keep in mind this: Any moderately or strongly supported hypothesis may be but a step away from one more piece of supporting evidence that will confirm it. Knowing this, you may be able to identify the kind of evidence it would take to "be that third leg of the stool," and then find out if such evidence exists. (Turner (2007) describes how an interlocking web of support has occurred in certain research in cosmology.)

4.8 Why researchers have to be good at using both the retroductive method and the H-D method

How would science progress if scientists could use only the retroductive method, or could use only the hypothetico-deductive method, or could, as they do, use both? To answer this, let's set time back to 1000 A.D. and run it forward in three separate worlds.

The first is the "retroduction world." If you are a scientist in it, you can't use the hypothetico-deductive method. You have one way for discovering knowledge and that is using the retroductive method to conceive hypotheses, and then supporting the hypotheses with circumstantial evidence and reason.

When this world reaches the present, Newton's theory of gravitation, and Darwin's theory of evolution have been, it's safe to say, reliably retroduced and supported. And scientists have reliably retroduced why the night sky is dark, and why snakes have forked tongues, and the list goes on and on. Yet we would lack much knowledge that exists in the actual world today. For instance, it's unlikely that in the retroduction world there would be knowledge of Dirac's equation. For, it's highly improbable that anyone could by looking at the fact that the spin of the electron is one-half, and the fact that antiparticles exist, retroductively come up with Dirac's equation, a mathematically abstruse reason that accounts for these facts.

The second world is the "hypothetico-deduction world." If you are a scientist in it, you can't use the retroductive method. You have to get hypotheses by the question method and the subject method. When the hypothetico-deduction world reaches the present day, Dirac's equation would have probably been discovered as Dirac discovered it by: (1) dreaming up mathematical axioms; (2) from them deducing the hypothesis, i.e., his equation; (3) from it deducing test predictions about the spin of the electron and about the existence of antiparticles; and (4) gathering facts and seeing they matched what was predicted. Likewise, millions of other discoveries would be made with the H-D method during the thousand years of this world.

The third world is best equipped for discovering. It's our world, where both methods are available to use separately and in tandem. We have a great deal of knowledge that was made reliable through hitching the two methods in sequence: (1) retroduction to generate a hypothesis, then supporting the hypothesis with circumstantial evidence, and (2) hypothetico-deduction to test the hypothesis. Such knowledge includes Einstein's two theories of relativity: Einstein retroduced both, and over the years others have corroborated them with a variety of H-D tests. Such knowledge would have been less reliably known had just one of the methods been used.

These three worlds as good as say, "Train yourself to be as expert as you can be in all the methods for conceiving and supporting hypotheses – the retroductive method, the question method, and the subject method – and in the method of testing hypotheses, the hypothetico-deductive method."

4.9 How to present H-D tests to an audience

Near the start of presenting your H-D research in a journal article or seminar talk, specify the hypothesis you tested and the test prediction(s) used. Near the end, say whether the test disproves or corroborates the hypothesis. Provided the test corroborates the hypothesis, say how certain you believe it is that the hypothesis is true. In detail now:

1. Specify the hypothesis tested and the test prediction(s) used. Near the start of your article or talk, tell the readers or listeners that you are reporting on a H-D test (or several). (I can't tell you how many times I have been unable to tell whether research reported in journal articles was based on the hypothetico-deductive method or instead on the retroductive method: the author(s) didn't say or give a clue.) Next, state the hypothesis and test prediction(s). I require the students in my Best Research Practices course to study Vander Wall's (1982) research article to see how he does this. On the article's first page, his first heading is "Hypotheses and Predictions." Under it he lists five hypotheses and their predictions. To show the form, let me quote the first three:

"**Hypothesis 1.** Nutcrackers find cached seeds using cues (e.g. olfactory) which emanate from the cache. Predictions: (1) birds should find each other's caches and those made by the experimenter; and (2) birds should not dig at a cache site where the seeds have been removed. Prediction (2) assumes no detectable residual cues remain at the cache site. This assumption will be justified below.

"**Hypothesis 2.** Nutcrackers find cached seeds using microtopographic features of the soil which are produced during cache preparation. Predictions: (1) birds should find each other's caches; and (2) if the microtopography of the soil is altered, caches cannot be located.

"**Hypothesis 3.** Nutcrackers recover seeds by random search. Predictions: (1) birds should find each other's caches and those made by the experimenter; and (2) success rates should be proportional to percentage of ground surface saturated with caches."

To recap, don't beat around the bush in telling readers or listeners what you did. Tell them right at the start. Do as Steve Vander Wall does.

As to the language for describing the H-D test(s), I recommend sticking with the standard terms, namely "research hypothesis" or "hypothesis," and "test prediction" or "prediction," rather than synonymous words. And don't shy from the terms "hypothetico-deductive method" or "hypothetico-deductive test." They are reminders of the method's integral features, such as clamping down on known and unknown shunt causes. And be sure to cite this book or another that explains the logic of the hypothetico-deductive method.

I'll tell you one place it's fine not to use the above terms and instead go with everyday language: abstracts and summaries of research. There, if you like, it's fine to use "view" or "conjecture" instead of "hypothesis," as with: "Our data supports the view (conjecture) that. . . ." Let me quote Cowen's (2006) perfectly clear abstract that doesn't have the words "hypothesis," "test prediction," and "hypothetico-deduction": "Some planetary scientists, including members of the Stardust team, had conjectured that many microscopic comet grains had formed near stars other than the sun and then entered the solar system during its youth. If so, the grains would contain a wide variety of isotopes from elements heavier than lithium. Instead [the investigators] found that the isotopic composition of nearly every grain that they analyzed matched that of the inner solar system. . . ." But apart from an abstract or summary, consistently use the standard terms in the body of a paper or seminar talk.

2. Say whether the test disproves or corroborates the hypothesis.

(1) If the test prediction is false, say something along the following lines: "The hypothesis that . . . is false." Or, "My research falsifies (disproves, contradicts, refutes, goes against) the hypothesis that. . . ." (Though you already know it, it's worth repeating that a hypothesis can't be moderately wrong, or strongly wrong, or very strongly wrong. If it's wrong, it's wrong.)

(2) If the test prediction is true, tell the audience: "The H-D test corroborates the hypothesis that. . . ." Or put it in other language that means the same. The point is that corroborate means to make more certain, to make us more willing to wager that the hypothesis is true.

3. Say how certain you believe it is that the hypothesis is true. H-D corroborations differ in their power to raise the percent certainty that the hypothesis is true. With a few exceptions, a H-D test with several test predictions, all which prove true, raises the percent certainty more than a H-D test with one test prediction that proves true. A H-D test designed to clamp down on known shunt causes so they cannot operate, and which incorporates a control group to rule out unknown shunt causes, raises the percent certainty nearly all the way.

At any rate suppose the H-D test has corroborated the hypothesis, by some degree made it more certain. Then, try to make a grand estimate of the likelihood that the hypothesis is true. The grand estimate considers all of the research that has been done, including yours, suggesting that the hypothesis is true. Lay all of this out for members of your audience so they know the bases of your belief.

One approach to making a grand estimate is to arrive at a "percent certainty" estimate that the hypothesis is true, reflecting your subjective feelings of probability. For instance, a 90 percent certain estimate interprets the feeling you would have in a gambling game in which there is a 90 percent chance of being right with a bet. Once you get an approximate percent certainty estimate, translate it into words that convey its essence.

Here, for your consideration, is how I do this:

(1) If I feel between 60 percent and 95 percent certain that the hypothesis is true, I say that the hypothesis is "suggestive rather than conclusive."

(2) If I feel 95 to 99 percent certain, I say that the hypothesis is "quite probably true," or I say it is "true beyond a reasonable doubt."

(3) If I feel more than 99 percent certain, I say that the hypothesis "is confirmed." Confirmed means firmly established. A confirmed hypothesis is true beyond virtually all doubt.

Don't feel bound to copy my language. There are various ways of saying, for example, that a hypothesis is suggestive rather than conclusive.

Expect some to disagree with your estimate of the likelihood that the hypothesis is true. Yours is a subjective estimate, theirs are subjective estimates. With time comes debate and sharing knowledge among scientists. The upshot is that some scientists re-estimate their estimates, and with time comes convergence to a consensus.

By regularly reading research reports in your field, you will develop a second sense for considering together all the research that favors a hypothesis, weighing the degree of corroboration each provides, and making an overall judgement of percent certainty that the hypothesis is true. And you will learn a variety of phrases for translating that percent certainty into language that accurately expresses it.

Comment: In convincing others of the degree of support that you believe a research project of yours provides to the hypothesis, the order in which you present the evidence matters. Logically it shouldn't, but in practice it does. This is why writers of essays, and trial lawyers, present the moderately supported points first, and the strongly supported last. It makes sense to do this in presenting research results too. Namely, to do the opposite, to give the strongest support first – that your H-D test corroborates the hypothesis – is apt to make researchers in the audience be less thorough in examining the bases of the weaker support, the circumstantial evidence and reason supporting the hypothesis.

If, so to speak, you hit 'em on the head with a sledgehammer first, they'll be insensible to all the little taps. Hence, give your audience all the little taps first, holding the sledgehammer for last.

4.10 What the word "theory" means

I'd like to end this chapter with the two meanings of the word "theory." Both are in wide circulation. Lay people often get them mixed up, and that spells trouble. In one, a theory is a hypothesis. In the other, a theory is a law.

When "theory" means "hypothesis." Scientists, in speaking of untested conjectures, sometimes use the word "theory" and sometimes the word "hypothesis" (short for "research hypothesis"). We hear physicists, for example, speak of "string theory." String theory is an untested conjecture of nuclear physics. Instead of "string theory" it would be equally correct for physicists to say "the string hypothesis." At the same time, we hear ecologists speak of the "Gaia hypothesis." The Gaia hypothesis, which casts Earth as a gigantic organism, is an untested conjecture of ecology. Instead of saying "the Gaia hypothesis," it would be correct to say "the Gaia theory."

When "theory" means "law." "Theory" means "law" whenever the theory referred to has been experimentally confirmed. During the early history of science, whenever a theory was confirmed, scientists no longer called it a theory; they called it a law. In modern times, with the advent of atomic physics, scientists stopped using the word "law." Whenever a theory becomes confirmed, scientists continue calling it a theory, but think of the word "confirmed" being tacitly attached to the word "theory." As Tyson (2003) notes, "modern theories are just as thoroughly tested and just as successful as the ideas that were previously known as laws."

Today we have Einstein's theory of general relativity instead of Einstein's law of general relativity. "The theory of general relativity," writes Tyson (2003), "didn't force us to discard Newton['s laws], but it encloses the phenomena of the universe that Newton described and encompasses other phenomena. . . ." More specifically, not only does Einstein's theory of general relativity correctly predict all the facts that Newton's laws correctly predict, Einstein's theory correctly predicts certain facts that Newton's laws are unable to predict accurately in some cases, and not at all in others. This doesn't mean that Newton's laws are wrong. Newton's laws are right for predicting facts within their domain.

Another example is the Big Bang theory of the origin of the universe. Tyson (2003) explains that the Big Bang theory "is supported by an overwhelming body of evidence. . . . The Big Bang is termed a theory [and not a law] in deference to the idea that it may someday be enclosed in a larger picture."

Another example is Darwin's theory of evolution. Its confirmation has come through thousands of retroductive researches based on a gradually gathered mass of factual and reasoned support concerning the fossil record and animal and plant breeding, all of which adds up to a smoking gun. It is a law within its domain of natural phenomena. But it isn't called a law because someday a new theory may be conceived and confirmed that explains all Darwin's theory does and explains more. So, when lay people say, " Darwin's theory is just a theory, and theories are only speculations," they wrongly take "theory" to mean "proposed theory" – a guess – and not the "confirmed theory" that Darwin's theory is. Had Darwin's theory of evolution been confirmed a hundred and twenty years ago, before the terminology shifted from "law" to "theory," rather then being confirmed after that, scientists at the time would have dubbed it "Darwin's law of evolution."

Coming full circle in this chapter. I opened this chapter by saying that although the hypothetico-deductive method was given its name in the middle of the 20th century, it has been practiced for hundreds of years. I will close this chapter by backing this statement up. Listen to Hedge (1884, p. 371) describe a hypothetico-deductive test of Copernicus' hypothesis that the planets circle the sun: "When Copernicus propounded the soli-central hypothesis, astronomers objected that, if his position were correct, Venus ought to have phases like the moon. Copernicus, nothing abashed, admitted the inference, but immediately added that if men should ever come to see Venus more distinctly, they would find that she had phases. This was before the invention of the telescope. When that instrument was given to science, one of its earliest fruits was the discovery of the phases of Venus."

5

How to discover cause and effect

I want to ask a simple question. By what methods can we discover cause and effect whenever it really exists? It turns out that there are two methods. Both involve construct X, the supposed cause, and construct Y, the supposed effect.

With the direct method, we take control. For n repeated experiments, we change the state of X and measure the effect on the state of Y. With the indirect method, we passively observe n cases of the states of X and Y changing on their own accord, and from the observations and other information we infer whether X is or is not causing Y. As to the proper value of n for statistical reliability, we join with a collaborating statistician to compute it from statistical theory (the two methods have separate equations for this).

Among the general ways of indicating "X causes Y" are: "X produces Y," "X determines Y," "X affects Y," "X influences Y," and "Y depends on X." Among the specific ways are: "X deters Y," "X impedes Y," "X stabilizes Y," "X increases Y," and "X reduces Y." Thus, a title of a research report, "Jute Matting Stabilizes Soil Loss," indicates a specific sort of cause-effect relation.

To say that X causes Y means that a change in the state of X causes, at least some of the time, a change in the state of Y. Just so, the heading "The Influence of Divorce on Depression" means that divorce can cause depression, not that it inevitably does. "Coffee Drinking Linked to Birth Defects" means that pregnant women who drink coffee are at increased risk of bearing a child with a birth defect(s). And "Women Who Consume Dairy are Five Times More Likely to Give Birth to Twins" obviously doesn't mean that consuming dairy products causes twin births in every case.

Another thing, although 19 of 20 cause-effect research projects are the kind where researchers hope to show that X causes Y, the less common kind is where they hope to show that X doesn't cause Y. Unrelatedness can be important. Think of Phase I clinical trials. They succeed when the novel drug being tested is shown not to cause a dangerous side effect(s). Or think of a study that brings peace of mind by discovering that a strange virus that attacks animals doesn't go after people.

As to nomenclature, when we start a project to see if cause and effect exists, it's proper to attach the word "supposed" to the words "cause" and "effect." This is because at the start we have a hunch – a shred of circumstantial evidence and/or reason – that a certain X causes a certain Y. If we then discover that X does cause Y, the word "supposed" is dropped.

This chapter has important messages about investigating for the possible presence of cause and effect. Section 5.1 covers the direct method, and section 5.2 covers the indirect method. Section 5.3 stresses the importance, when applying either method, of minimizing systematic error and random error. Section 5.4 explains how the indirect method has a good deal going for it: it is responsible for much of science's important knowledge of cause and effect. Section 5.5 cautions to be on guard against pseudoscientifically derived claims of cause and effect. And last, section 5.6 presents a brief description of reciprocal cause-effect relations.

5.1 The direct method of discovering cause and effect

With the direct method, each of its *n* trials goes this way: We first measure the state of X, the supposed cause, and Y, the supposed effect. Then we apply a treatment to X, changing its state. After a suitable delay we measure the state of Y. If it has changed, probably the change in X caused it to change.

Less often we cannot change X but some agency does it for us. Sweden provides a good example. At one time, Swedish law had everyone driving on the left-hand side of roads. The law changed on September 3, 1967, specifying that everyone drive on the right-hand side. With the direct method, researchers investigated the effect of the new law – a treatment, so to speak – on traffic accidents. (The example incidently illustrates an experiment limited to one trial; the treatment was applied just once.)

The idea of "compared to what" is central for judging the degree to which X causes Y. The "what" is a baseline, also called a comparison group. There are four kinds of baselines in all. To explain them, suppose the treatment is a new drug for a certain disease, and it's being tested in a clinical trial.

No-treatment baseline. The no-treatment baseline is the average state of Y that results when no treatment is given. In terms of the new drug, the average state of the disease in the patients could get worse or improve without treatment. "Of those taking the drug, 17% went into remission" sounds good until we learn, "Of those taking nothing, 24 percent went into spontaneous remission."

Before-treatment baseline. The before-treatment baseline is the state of Y before the treatment is given. Concerning the drug, we measure the average state of the disease in patients before they receive the drug, giving the before-treatment baseline. Any improvement or not from receiving the drug is judged against the before-treatment baseline. "The average total cholesterol of patients who received the drug was 217 ml/d at the end of the study" indicates nothing about the effect of the drug unless we know the average at the beginning of the study, before the treatment was given.

Blank-treatment baseline. The blank-treatment baseline is the state of Y that results from a blank treatment. Other names for a blank treatment are dummy treatment, inert treatment, or placebo. A blank treatment is missing the active substance of the actual treatment, but in all other respects is the same. For the drug test, we give a randomly selected group of patients a placebo, and give the rest the actual drug. Neither researcher(s) nor patients know who got what, drug or placebo (called a double-blind, placebo-controlled study). Any change in the disease from treating patients with the drug is judged against the change from treating patients with the placebo.

Existing-treatment baseline. The existing-treatment baseline is the state of Y that results when the existing treatment is applied. If there is an existing drug for the disease, is the new drug more effective than it? To answer this, we have to know the existing-treatment baseline.

Not every one of the four baselines may be possible. Suppose that X is a novel teaching method for raising children's appreciation of nature, Y. There is not a blank-treatment method for raising children's appreciation of nature. It would have to be a method for teaching nothing, and there is no such thing. Or suppose we are investigating cause-effect in basic science. In basic science, usually there is no existing treatment. X is changed simply to learn if it produces a change in Y, and how much the change is.

Res. Ex. 5.1: Does alcohol stimulate the growth of cancer?

I have before me a research report titled "Ethanol Stimulates Tumor Progression and Expression of Vascular Endothelial Growth Factor in Chick Embryos," by Gu et al. (2005). It begins with a list of various epidemiological observations suggesting that drinking alcoholic beverages is a risk factor for human cancers, including those of the mouth, larynx, esophagus, breast, liver, and large intestine. Because ethics prohibits planting cancers in people, feeding half alcohol and half not, and seeing if alcohol has an effect on tumor growth, Gu et al. approximated that. They planted cancerous tumor cells in 12 chick embryos. For the actual treatment, they injected in each of six of them 0.25g/kg per day of ethanol. For a blank treatment, in each of the other six they injected 0.25g/kg per day of saline solution. Later they discovered that the ethanol-fed tumors had grown nearly four times as much as the saline-fed. By analogy, they concluded that people with cancer who drink alcohol may risk increased tumor growth.

For the comparison group – the blank treatment baseline – why didn't Gu et al. instead take six embryos with implanted tumor cells and leave it at that? Why inject them with an inert substance, a saline solution? Injecting an inert substance rules out the possibility that foreign material – which ethanol is, apart from its particular chemical structure – could irritate the tumors, stimulating or inhibiting their growth.

Always in applying the direct method, we make sure the sample units (chick embryos in the above example) that we treat have the exact same properties. This is to rule out two possibilities: (1) the possibility that one or more sample units have properties that by themselves affect Y, while the rest don't; (2) the possibility that one or more sample units have properties that strengthen or weaken any effect that X has on Y.

Res. Ex. 5.2: Do journal referees discriminate against authors?

In case you believe researchers are untouched by name dropping, read on. It just might be that having their names and affiliations on research articles they submit to journals affects the chance of being published. Blank (1991) decided to see. She worked with the editors of *The American Economic Review* for a period in which they sent 1,498 submitted articles to reviewers. The editors randomly chose 832 of the articles and gave them a novel treatment: they removed the authors' names and affiliations. On the remaining 666 articles they let the names and affiliations appear, as had been the journal's practice.

Blank found that X – withholding name and affiliation – lowered Y, the chance of acceptance. The reason, she retroductively surmised, is that reviewers judge the articles more critically when they have no idea who the author(s) is.

Res. Ex. 5.3: Do antipollution laws have an effect?

New environmental laws are enacted, limiting the amount of pollutants discharged from factories. Do good environmental consequences follow? Specifically, in the decade before 1970, lead concentrations in fresh Greenland snow were more than 200 times greater than in snow buried a thousand years ago. In 1970, the United States and other countries enacted laws limiting lead in gasoline. Twenty years later, Boutron et al. (1991) collected and analyzed samples of Greenland snow. The concentration of lead averaged 7.5 times less than in 1970, with similar decreases of cadmium and zinc. The antipollution laws, X, very probably caused pollutants, Y, to decline. It's unimaginable that anything else could have.

Res. Ex. 5.4: How did James Lind pinpoint the cause of scurvy?

The symptoms of scurvy include pain in the joints, weak knees, putrid gums, black-and-blue marks on the skin, and death. Outbreaks had long been noticed among sailors confined at sea, and the cause was unknown until James Lind, physician on the HMS Salisbury, performed six cause-effect experiments. Suspecting that an incomplete diet was at the root of it, Lind (1753) selected 12 crew members with symptoms, divided them into six sets of two, and gave them daily treatments in addition to their normal meals. Set 1 got a quart of apple cider; set 2 got 25 drops of elixir vitriol; set 3, six spoonfuls of vinegar; set 4, half a pint of sea water; set 5, two oranges and a lemon; and set 6, a mixture of spices including nutmeg, garlic, mustard seed, and gum myrrh. The symptoms went entirely away in those who got the citrus fruit, partly away in those who got the cider, and the rest failed to improve. Consequently the British Royal Navy began giving its sailors daily rations of the juice of limes and lemons (hence the name, limeys).

As Lind's research reminds us, when trial and error hits upon a treatment that works, it raises the question of the most effective dosage. The answer comes through a dose-response study. A range of treatment levels, different strengths of dose, is tried with groups of patients to see which works best. For more on dose-response research see Cox (1968) or most any book on clinical trials.

And as Lind's research also reminds us, after X is found to cause Y (whether by the direct or the indirect method), attention moves to trying to explain the causative mechanism. Concerning scurvy, knowledge of chemistry grew to where in 1932 it allowed lime juice's preventive ingredient, vitamin C, to be identified. Today, the molecular mechanism of how vitamin C operates on the body's collagen fibers, keeping them healthy and scurvy away, is understood.

I'll tell you two more examples of attention moving to trying to explain a causative mechanism.

Ex.: It was accidently discovered that putting salt on icy surfaces in winter melts the ice. That aroused interest in knowing why, which led to research in salt chemistry that revealed the molecular process of the melting.

Ex.: Dermatological researchers discovered that a drug called Restylane, when injected in adults' faces, smooths out skin wrinkles. Next, Wang et al. (2007) disproved one hypothesis of how Restylane works and confirmed another. Disproved was that Restylane fills the space left by collagen loss in the skin. Confirmed was that Restylane causes a biological response, reforming the aged collagen to its youthful condition. A report on this in *The Wall Street Journal* (2/20/07) was headlined "Wrinkle Drug's Action Is Found." Appearing in news releases were two relevant comments. David J. Leffell, Yale University School of Medicine said: "[The research is important] because it begins to build a scientific basis for a cosmetic procedure that has historically just been based on the end result." June Robinson, editor of *Archives of Dermatology*, said: "[The study] is the definitive paper that tells us how this product works."

Res. Ex. 5.5: Is it wise to go into surgery relaxed?

Relaxed would be better than uptight, wouldn't you think? And until recently physicians thought the same. For a long time they'd known that high levels of stress hormones and blood pressure in patients undergoing surgery contribute to postoperative fatigue, and to impairment of the immune system. Common sense told them that relaxation could help keep the levels down. The question was, how much down?

Manyande et al. (1992) decided to see. They randomly assigned to a treatment group or to a blank-treatment group each of 40 patients coming into a hospital for minor surgery the next day. Before their surgery, those in the treatment group listened to a 15-minute tape recording that was designed to relax them: "Think about the temperature of your hands . . . and any tingling feelings in your fingers and hands. You may notice slight tension or tightness around your wrists. . . . As you notice the tension, let your hands go heavy and limp. . . ." Patients in the blank-treatment group listened to a 15-minute "placebo tape" about the hospital and its staff, containing information like: "The hospital was founded in 1835. . . . Your care is in the hands of a team of doctors that is led by a consultant surgeon. From day to day you will be looked after by registrars. . . ."

During surgery, the researchers measured the levels of stress hormones in each patient. Contrary to expectation, the patients who listened to the relaxation tape had higher levels of stress hormones than those who listened to the placebo tape. A possible explanation, offered Manyande et al., is psychologist Irving Janis' concept that anxiety benefits people by keeping potentially harmful stress down. Relaxation, by suppressing anxiety, may allow stress to rise.

Res. Ex. 5.6: Does heart bypass surgery cause permanent memory loss?

Most heart patients lose some of their memory after bypass surgery. Among the theories retroduced to explain this are: (1) the surgery releases tiny particles that travel to the brain, causing ministrokes; (2) blood clots and bubbles form in the heart-lung machines during the surgery and get lodged in the brain; (3) anesthesia, besides putting patients to sleep, produces changes in the brain, affecting memory.

Years after bypass surgery, might memory return to its pre-surgery level? To answer this, Newman et al. (2001) established a before-treatment baseline (pre-surgery): they measured with standard memory tests the memories of 261 patients before they had bypass surgery. Later, they measured the memories of the surviving patients at discharge from the hospital, again at six months, and again at five years after discharge. Relative to the before-treatment baseline, memory at discharge was 53 percent lower. At six months it was 25 percent lower, having recovered some.

The unexpected finding was that at five years the decline had worsened; memory averaged 42 percent lower than the pre-surgery level, nearly what it had been at the time of discharge.

Lack of a no-treatment baseline is partly why Newman et al. call their findings preliminary. At this writing, a study announced by Selnes and McKhann (2001) is underway to compare the memory loss of those who have undergone bypass surgery with those with similar cardiovascular risk factors and symptoms who haven't.

Res. Ex. 5.7: Does acid rain reduce bird populations?

At the start of their research, Graveland et al. (1994) knew the following: In The Netherlands 80% of the forests are on poor sandy soil, and the soil was becoming increasingly acidified by acid rain. In these forests, the proportion of great tits laying eggs with defective shells increased from 10% in 1983-84 to 40% in 1987-88, with similar increases noted for other passerines. The researchers decided against DDT causing the defective shells. For one thing, DDT had long been banned in The Netherlands. For another, the eggshells of birds of prey in The Netherlands, which are highly vulnerable to DDT-caused thinning, had normal shell thickness.

Graveland et al. hypothesized the following chain of cause and effect: acidic soils (a state of construct A) cause snail populations to decline (a state of construct B); which limits the number of snails shells (a state of construct C) that egg-laying great tits eat; which leaves their diets deficient in calcium (a state of construct D); which makes the shells of their eggs thin (a state of construct E), increasing breakage; which reduces the number of great tits born (a state of construct F). Accordingly, with separate cause-effect studies the researchers corroborated the key empirical links of the chain:

(1) They showed that acid soils had caused the number of snails to decline. To wit, they added lime to the soil of some forested plots, which decreased acidity, and left nearby control plots be. They found that the snail populations recovered in the treated plots but not in the control plots. Besides, these results were supported by results of a previous study in Swedish forests.

(2) They showed that female great tits depend on snails for their major source of calcium. When they made snail shells scarce, the birds took to scavenging the shells of chicken eggs from picnic sites and farms.

(3) They showed that lack of calcium causes the shells to break. To do this, they fed a group of birds snail shells in feeders attached to their nest boxes, while not feeding shells to a comparison group. The birds fed calcium produced normal eggs; the birds not fed calcium produced a good many defective eggs.

The research, said ecologist Kenneth V. Rosenberg, of Cornell University, "illustrates what to most people would be an unexpected link between acid rain and bird populations. It's something that [without research] could go completely unnoticed."

5.2 The indirect method of discovering cause and effect

The indirect method is practiced when neither researchers nor an agency can make X happen on command. X simply happens, and after that a supposed effect Y sometimes happens. But is X causing Y to happen? Or are their happenings only correlated? To answer that, researchers follow the four steps of the indirect method. Steps 1 and 2 involve testing a statistical hypothesis. Steps 3 and 4 involve scientific detective work.

Step 1: Make measurements. Take a random sample of n sample units – people, plots of land. . . . On each measure the states of the main constructs, X and Y. Also measure the states of any incidental constructs. An **incidental construct** is one other than X, where a change in the state of the incidental construct can, possibly by itself or possibly by interacting with the state of X, affect the state of Y.

Step 2: Do a statistical analysis of the measurements. In the statistical analysis (e.g., a regression analysis), test the statistical hypothesis that X and Y are uncorrelated, while having the statistical analysis control for the effects of any incidental constructs. In the event the test leads you to decide that X and Y are uncorrelated (as portrayed in Fig. 5.1a), that means X is not causing Y, and Y is not causing X. In the event it leads you to decide that X and Y are correlated (as portrayed in Fig. 5.1b), continue to step 3.

Step 3: Determine whether or not some construct Z is making X and Y be correlated. Determine whether there is a construct Z that is both a cause of X and a cause of Y, i.e., simultaneously driving the states of X and the states of Y, making them correlated (as portrayed in Fig. 5.1c). If you find such a construct, conclude that X is unlikely to be causing Y, nor Y causing X. On the other hand, if you are confidently able to rule out the existence of a third construct Z, continue to step 4.

Step 4: Try to produce a convincing argument that the correlation of X and Y is because X causes Y (as portrayed in Fig. 5.1d). Gather any of four kinds of supplemental information for the argument. (1) Gather **primary evidence**: evidence from as diverse a lot of past researches (done before yours) as is available that show a correlation between X and Y. (2) Gather **secondary evidence**: evidence from a diverse lot of past researches that show correlations between constructs similar to X and Y. (3) Gather any logical reasons for why Y probably does not cause X. (4) Gather any logical reasons for why X probably does cause Y.

The order in which the four are listed has no relevance. Work on them together if you like.

Concerning steps 2 and 3, the indirect method rests on the logic that correlation is necessary for causation, but is insufficient. So if X and Y are uncorrelated, then X does not cause Y, and that's the end of it. But if there is a correlation between X and Y, it could result from X causing Y, from Y causing X, or from some construct, Z, causing both X and Y. To illustrate the latter, an article in *Science News* (1 May 2004), titled "Unsettling Association: Dental X rays linked to low-birth-weight babies," presents research results suggesting that pregnant women who had their teeth X-rayed were more likely to bear underweight babies. In response, a reader (Stephen Wood) wrote a letter published in *Science News* (3 July 2004) saying: "Here is an alternative possibility: Perhaps unhealthy people are more likely to have low-birth-weight babies and bad teeth."

Usually, though not invariably, with the indirect method a conclusive case for cause-effect cannot be built in a single go. Each study in favor of cause-effect, with none against, inches the community of researchers nearer to the day it wakes up and realizes that the accumulation of studies has crossed a tipping point, and that "X causes Y" is confirmed.

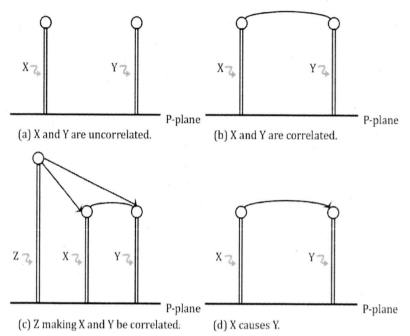

Figure 5.1. P-plane diagram depicting (*a*) X and Y uncorrelated, (*b*) X and Y correlated, (*c*) Z causing both X and Y, making them correlated, and (*d*) X causing Y.

Res. Ex. 5.8: How do we know that smoking causes lung cancer?

Let's look at how the medical research community went from being unsure that smoking cigarettes causes lung cancer to being very sure. A mass of information, mostly from research projects based on the indirect method of studying cause-effect, gradually and steadily was accumulated to where it crossed a tipping point into belief that smoking certainly does cause lung cancer.

First, with the years came a great collection of primary evidence from a diverse lot of research showing that smoking and lung cancer were correlated. By "a diverse lot" I mean the studies involved diverse background conditions, including where the people lived, their eating habits, their incomes, their kinds of work, their stress levels, their genetic profiles, and so forth. The invariance of the correlation to the diverse background conditions meant there was next to no chance that a third construct, Z, was behind the correlation, because the odds are against any such construct being common to all the studies.

Besides, no construct Z was imaginable that could drive both X and Y. Any such construct would have to affect a behavioral condition, smoking, and a physical condition, lung cancer. And since lung cancer appeared after smoking, lung cancer could not cause smoking.

Second, with the years came a great accumulation of diverse secondary evidence that supported the idea of cause-effect. Epidemiological research found that cigarette smoke is carcinogenic to lung tissues of laboratory animals. It found that pipe smoking and cigar smoking are associated with chronic coughing, and with cancers of the head, neck, and lungs. It found that use of moist snuff is associated with cancer of the jaw. It found that dogs living indoors with smokers are more likely to develop lung cancer than those living indoors with nonsmokers, and specifically short-nose dogs that are habitually around smokers are especially susceptible to lung cancer (the nose acts as a filter). Added to this was a big pool of anecdotal evidence. Most of us have heard talk along the lines of uncle Ted's chronic smoking and coughing ended with him dying of lung cancer at 46, while his twin uncle Bob never smoked, is in the pink, and is now 87.

Third, it appeared as if "the more the cause, the more the effect": heavy smokers seemed more likely to get lung cancer than did light smokers. And as far back as the late 1800s a slang word for cigarettes emerged from the anecdotal pool: coffin nails.

Fourth, search as they did for contrary evidence, researchers came up empty handed. No conceivable reason or known evidence supported any notion of health benefits from smoking.

Think of the power of conviction that comes from a diverse lot of secondary studies involving correlations. For instance, take the idea that stress or a genetic predisposition (Z) might make people smoke (X) and also make them get lung cancer (Y). The studies with pet dogs go against that idea: disproportionately more dogs that lived with smokers got lung cancer than those that lived with non-smokers, even though dogs are neither commonly stressed nor are their genes entirely those of people. On the power of conviction that comes from secondary studies, i.e., a diverse lot of studies that show correlations between constructs similar to X and Y, hear Cox (1968): "The plausibility of an association's having a cause-effect relation increases with the number of different and unrelated sources from which like or supportive information is derived. . . . The hypothesis that lung cancer is causally associated with cigarette smoking is supported from associations of per capita cigarette consumption with mortality, from sex differences in mortality correlating with cigarette consumption, from dose-response relations rising with consumption and falling with cessation, from human histologic findings of premalignant changes in lungs of smokers and of changes attributable to smoking in patients with lung cancer, and from animal experimental evidence of seven extractable carcinogens from tobacco smoke. . . . Supporting evidence coming from different methods of study provides coherence, binding together the parts with a web of interdependent threads."

Go to the original research reports on which the following research examples are based and you will see that the researchers practiced the four steps of the indirect method. The research examples abridge the originals and give only highlights.

Res. Ex. 5.9: Can newspaper stories cause airplane crashes?

From studying newspaper stories in metropolitan regions, Phillips (1978) found a correlation: often within a week after stories of murder or suicide appeared, one or several fatal crashes of private airplanes happened. Whereas during periods without such stories, fatal crashes were fewer. Phillips decided that cause-effect might be behind the correlation, and it was impossible that fatal crashes could cause just-previous murders or suicides not involving crashes. Phillips concluded: "The evidence thus suggests that (1) some persons are prompted by newspaper stories to commit murder as well as suicide, and (2) noncommercial airplanes are sometimes used as instruments of murder and suicide." Retroductively, he went on to propose possible motives: wanting to disguise suicide to maintain one's honor, or to disguise murder or suicide to keep a life insurance policy valid.

The direct method is powerless to reach this conclusion. Researchers would have to make suicides and murders happen and then monitor for spikes in fatal crashes.

Res. Ex. 5.10: Do forest fires increase the level of mercury in fish?

In designing an experiment to answer this, Garcia and Carignan's (2005) thoughts ran this way: After forests in the Canadian Boreal Shield have been clear-cut or partially burned, mercury (Hg) in the topsoil is possibly released. Washed into streams, it would get into plankton, be carried in them downstream into lakes, and some plankton with mercury would be eaten by fish belonging to the lowest trophic level. Fish at the next higher level would eat some of those fish, and the fish at the next higher level eat some of those, and thus up the food chain to top predatory species – northern pike, walleye, and burbot – the mercury would become more concentrated.

Garcia and Carignan did three comparative cause-effect experiments. One involved lakes below partially burned forested areas, one involved lakes below clear-cut forested areas, and one involved lakes below undisturbed forested areas. For each of the three kinds of lakes they computed the correlation between mercury concentration and trophic level. Compared to the value of the correlation for lakes below undisturbed areas, the values were higher for lakes below clear-cut areas, and for lakes below partially burned areas. Suggesting cause-effect was at work, they wrote: "Thus, high Hg concentrations in fish from forest-harvested and partially burnt lakes may reflect increased exposure to Hg relative to that in lakes not having these watershed disturbances."

Note that X, the cause, is measured as a qualitative class, viz. the type of area – (1) partially burned, (2) clear-cut, (3) undisturbed. The question is whether or not X is correlated with Y, where Y itself is a correlation of Hg concentration and trophic level.

To do this research with the direct method would have required simultaneously clear-cutting randomly selected parts of watersheds, and partially burning randomly selected parts of others. As it was, the researchers had no control in applying treatments; they were applied by forestry interests. The researchers came in later and did an indirect cause-effect study concerning mercury concentration.

Res. Ex. 5.11: Is the insula the brain's organ of addiction?

It's not farfetched to think that the research of Naqvi et al. (2007) has started a chain of discovery that in time will help prevent addictions. They studied 32 former cigarette smokers, all with accidental injury to their brains. MRI scans of 16 of them showed damage to the insula, a region of the brain. After the damage had occurred, the 16 had lost their craving to smoke and easily quit. For the other 16 – a comparison group giving a blank-treatment baseline – MRI scans showed that their brain damage left their insulas intact. They retained their craving to smoke, and it was harder for them to quit.

In terms of constructs, X is insula damage, measured as present or absent; Y is difficulty of quitting smoking, measured as easy or hard. X and Y are correlated, a sign of possible cause-effect, and several ideas suggest it is. One is that injury to the insula came before the craving to smoke disappeared, rather than the reverse. Another is that researchers believe that the role of the insula generally concerns the hold of our bodily needs and emotions on us.

After concluding cause-effect, Naqvi et al. retroductively proposed several hypotheses for how the insula could be involved in all kinds of addiction, including drinking alcoholic beverages and eating chocolate. Awaiting is the next stage of other researchers proposing other hypotheses, and the stage after that, testing the best circumstantially supported hypotheses with the H-D method.

Again, the direct method is powerless to reach this conclusion. Researchers would have to randomly damage people's insulas, and see what if any effect that had on addictions.

Res. Ex. 5.12: Are genes an influence on macular degeneration?

Environmental factors, defective genes, or an interaction of the two – what is the cause of macular degeneration? Stone et al. (1992) decided to investigate. They studied 45 members of a five-generation family that was prone to having Best's disease, a form of macular degeneration. With ophthalmoscopy and electro-oculography analyses, they found that 29 had the disease and 16 did not. From DNA samples of the 45, they found a statistically significant correlation between X, having a certain defective gene, and Y, having macular degeneration. Where, as in this case, the disease-causing gene and the disease are each rare, and their occurrence is usually together, the association implies cause-effect. Hear Cox (1968, p. 7): "When two events are relatively uncommon, their frequent concurrence implies a causal relation."

Res. Ex. 5.13: Does time spent in school increase IQ test scores?

A review article by Ceci (1991) presents the agglomerative results of nearly 200 independent researches that found that the number of years students spend in school is correlated with their IQ level. Because the correlation has been found again and again in various schools, using research studies with different designs, the plausibility is strengthened that time spent in school causes the rise in IQ. Further, Ceci gives logical reasons for why it should, including: (1) Schools convey benefits on IQ test performance through the direct inculcation of relevant information. (2) Schools influence IQ test performance indirectly by inculcating modes of thinking and reasoning that are valued on IQ tests. (3) It is in schools that students first encounter the use of hypotheticals, and where learning and remembering are introduced as ends in themselves. (4) Schools foster an appreciation of hierarchical organization, an element of IQ.

Researchers trying to answer with the direct method the question, "Does time spent in school increase IQ test scores?," would have an impossible time of it. They would have to randomly select a group of students and have them spend little time in school, and randomly select another group and have them spend much time in school, and measure any resulting IQ differences.

Res. Ex. 5.14: Do pets make the elderly healthy?

Think of it! National health care costs going through the roof, and the solution may be for elderly people to own a dog or cat. To decide whether this cause-effect notion is true, Siegel (1990) conducted interviews with 345 randomly selected elderly pet owners and 593 elderly who had no pets. She collected data on the two main constructs: X, pet ownership, measured as "yes" or "no," and Y, doctor visits, measured as the number of meetings with doctors during the past year. Along with this she collected data on a number of incidental constructs, i.e., those other than X that could conceivably influence Y, doctor visits.

She entered all the data into a regression analysis program. It controlled for incidental constructs that could influence Y. She found a statistically significant correlation between pet ownership, X, and doctor visits, Y. She concluded that when "[the incidental constructs of] sex, age, race, education, income, employment status, social network involvement, and chronic health problems were controlled for, respondents with pets reported fewer doctor contacts during the year than those without pets." For instance, controlling for health problems went a good way toward ruling out the possibility that a construct Z – level of health – was driving both X – the choice or not keep a pet – and Y, the frequency of visiting doctors.

Siegel reasoned that cause-effect is behind the correlation of X and Y. Specifically, she reasoned that the elderly have need for attention. Those with pets tend to get it from their pets; some of those without pets schedule visits with their doctors as much for attention as for advice on health problems. "Altogether, these data," she says, "indicate that owning a pet, particularly a dog, may reduce the demand for physician services among the elderly. As all analyses controlled for health status, it appears that pet ownership is primarily influencing social and psychological processes rather than physical health. . . . Further support of this notion comes from data indicating that pet ownership reduces demand for care in times of stress. This latter finding is consistent with the growing literature on the role of social support in buffering the potentially negative consequences of stressful life events. It has been observed that only those social relationships that provide appropriate forms of support can act as effective buffers. Accordingly, dogs more than other pets provided their owners with companionship and with an object of attachment."

Once more, the direct method is powerless to reach this conclusion. Imagine, randomly select a group of elderly people without pets and randomly place pets with some of them. Then see whether or not those with the pets stop going to the doctor's office as often as they did before getting pets. The trouble with this is that giving pets to people who don't want pets would disrupt their lives. What would be discovered in this artificial, almost-laboratory situation would be inapplicable to the natural social world.

Res. Ex. 5.15: Do violent TV programs promote aggression?

Here is an application of the indirect method that isn't the least bit afraid to say, "I'm pretty convincing." It begins with Johnson et al. (2002a) asking, "Do youngsters who watch lots of violent TV shows later tend to become aggressive adults?" To answer this they randomly selected 707 Americans about age ten. During the next 17 years they interviewed the subjects' families four times, three to eight years apart. They measured X, the number of hours of violent TV each subject watched over the years. They measured Y, the frequency of each subject's later involvement in aggressive acts, such as assaults, robberies, and crimes committed with weapons. And they measured incidental constructs known to be statistically associated with aggression: each subject's sex, family income, childhood neglect, psychiatric disorders, and more. With a regression analysis of the data, controlling for the incidental constructs, they discovered a sizeable and statistically significant correlation between the amount of violent TV watched and later aggressive actions.

A correlation, yes, but was there cause and effect? What of the possibility that a construct Z that wasn't controlled for – say a constitutional predisposition for aggressive acts – might influence both X and Y, drawing subjects to watch violent TV and later be aggressive? Johnson et al. (2002b) thought this doubtful, saying "considerable evidence from other studies indicates that there is a bidirectional association between media violence and aggressive behavior."

Johnson et al.'s research adds to prior researches on the topic, all suggesting the same conclusion. It consists of more than 200 projects involving more than 50,000 participants (Anderson and Bushman 2002). These projects, from varied perspectives, found associations between exposure to TV violence in youth and later aggressive behavior. No dozen of these studies is sufficient to make us believe that removing violence from TV will promote a less aggressive society. But the totality is pretty convincing.

Think of trying to do this research with the direct method. Think of one large project, thousands of youngsters, half made to watch violent TV shows, the others kept from watching it. Any guess at the number of lawsuits?

Comment: Where X and Y are qualitative, each with two states, e.g., presence/absence, a simple way of exploring for cause-effect with the indirect method involves a 2 x 2 table, shown as table *a*, Fig. 5.2. Chitty (1967) has explained and illustrated it. His symbol for X is C, standing for the supposed cause, and for Y is E, standing for the supposed effect. When the supposed cause is present, we indicate its state as C; when absent, we indicate its state as ~C. When the supposed effect is present, we indicate its state as E; when absent, as ~E.

Here's a fictitious but illustrious application. We randomly select 34 people (a proper sample size for making an inference to the larger population of people, calculated with the help of a collaborating statistician, we are assuming for the sake of discussion). For each of the 34, we determine whether or not the person ate sweets over the Christmas holidays, and whether or not the person gained weight. C is "eats sweets," ~C is "does not"; E is "gained weight," ~E is "did not." Suppose the 34 counts fall in the four cells of the table *b*, Fig. 5.2. This would corroborate the idea that C causes E, i.e., that eating sweets causes a gain in weight. This is because of the 0 counts in the cell C•~E. All 10 who ate sweets are in the cell C•E. None are in the cell C•~E. It doesn't matter which cells of the bottom row the other 24 who didn't eat sweets are in. For the record, 8 are in cell ~C•E – didn't eat sweets, gained weight – and 16 are in cell ~C•~E – didn't eat sweets, didn't gain weight. Hence the data in the table lend support to the idea that eating sweets causes a gain of weight, or equivalently that eating sweets is sufficient but not necessary for gaining weight.

In the assumed event that the counts are as shown in table c, Fig. 5.2, with a zero in the top-right and bottom-left cells, that would suggest that eating sweets causes a gain of weight, and gaining weight only comes from eating sweets; i.e., eating sweets is necessary and sufficient for gaining weight. Such one-to-one correspondence of cause with effect rarely occurs in research; nature tends to have one or more causes of each effect, and one or more effects of each cause.

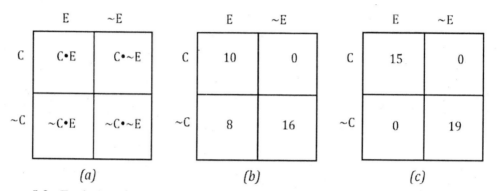

Figure 5.2. Exploring for cause-effect with the indirect method, for the case of two-state qualitative constructs, e.g., presence/absence. Table a is the general 2 x 2 table. C indicates the presence of the supposed cause; ~C indicates its absence. E indicates the presence of the supposed effect; ~E, its absence. The four cells indicate joint conditions. For example, ~C•E is the condition where the supposed cause is absent and the supposed effect is present. Tables b and c are for the example about eating sweets, described in the text.

It might happen that there is a zero count in one of the cells other than the top-right one. What then? Transpose the table's rows and/or columns to bring the zero to the top-right cell, which will indicate what is causing what. Fig. 5.3 demonstrates, where people are the subjects; the supposed cause is their exposure to noise, and the supposed effect is their being stressed. Let C be "noise present," ~C be "noise absent," E be "stress present," and ~E be "stress absent." We collect counts, and suppose they give the table a in Fig. 5.3. Next we interchange the two rows, carrying along their designations, C and ~C (table b). Finally, we interchange the two columns, carrying along their designations, E and ~E, to get table c, with the zero in the top-right corner. This table suggests that ~C causes ~E, i.e., absence of noise causes absence of stress.

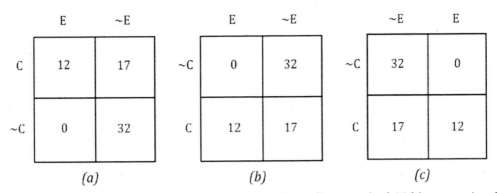

Figure 5.3. Example of exploring for cause-effect with the indirect method. Table *a* contains the original data. Table *b* has the two rows of table *a* interchanged. Table *c* has the two columns of table *b* interchanged, suggesting that ~C causes ~E, i.e., absence of noise causes absence of stress.

Comment: You open a research journal. You see a correlation analysis, e.g., a regression analysis. Don't immediately think the author(s) is investigating cause-effect. Correlations allow predicting Y from X even when X doesn't cause Y. Here are three illustrations:

(1) What is this: Y = 42.5 + 7.2X? It's an equation developed with regression analysis by fitting pairs of data – (x_1, y_1), (x_2, y_2), . . . , (x_n, y_n) – on X and Y for Yellowstone National Park's Old Faithful Geyser. X is the number of minutes it has just erupted; Y is the minutes wait to the next eruption. If, say, an eruption lasts X = 2.8 minutes, the equation predicts the next eruption will occur in Y = 42.5 + (7.2)(2.8) = 62.7 minutes. Evidently X is not causing Y, and Y is not causing X. Z, an unmeasurable process of refilling Old Faithful's chambers, determines both X and Y, making them correlated. Weisberg (1980) developed the equation from 1978's data. Park Rangers regularly consult an updated version of the equation to tell visitors the time the next eruption is expected.

(2) Physicians take advantage of correlations. I am thinking particularly of the research article of Wannamethee et al. (2006): "Height Loss in Older Men: Associations With Total Mortality and Incidence of Cardiovascular Disease." From part of the article's abstract: "A prospective study was performed on 4213 men whose height was measured between the ages of 40 and 59 years and again 20 years later between the ages of 60 and 79 years. . . . Height loss correlated significantly with initial age (r = 0.20) and weight loss (r = 0.20). Total mortality risk was higher in men with a height loss of 3 cm or more than in men with a height loss of less than 1 cm. . . . The excess deaths were largely attributable to cardiovascular and respiratory conditions and other causes but not to cancer." Despite no known reason for the correlation of height loss and mortality, the correlation can be helpful. It puts physicians on alert to watch out for height loss in older male patients, a red flag for possible onset of cardiovascular and respiratory problems.

(3) Police take advantage of correlations. Recently I read an article headlined: "Police capitalize on common failing of criminals: they break traffic laws. Minor traffic offenses are correlated with major criminal activity." The article quoted Charles Bahn, professor at the John Jay College of Criminal Justice in New York, saying: "People who are up to nefarious things consider it inconsistent to obey traffic laws."

5.3 The importance of minimizing systematic and random errors in cause-effect research

Recall from chapter 3 that research hypotheses, when retroduced from facts contaminated with sizeable systematic and/or random errors, run a sizeable risk of being wrong. Recall from chapter 4 that H-D tests performed with facts contaminated with sizeable systematic and/or random errors run a sizeable risk of being wrong. We are now in chapter 5, and guess what? In testing for cause-effect, if the main constructs, X and Y, as well as incidental constructs when using the indirect method, are contaminated with sizeable systematic and/or random errors, testing for cause-effect runs a sizeable risk of being wrong. Two examples will make this clear:

(1) Commenting in the *Archives of Internal Medicine* about a certain cause-effect study previously published there, Andrade (2006) notes the possibility that systematic error biased the results. His opening sentences tell the story: "[The authors] conducted a randomized controlled trial of transcendental meditation™ vs. health education to determine the benefits of meditation on components of the metabolic syndrome in patients with coronary heart disease (CHD). The authors correctly identified a number of limitations of their study. However, I wish to point out an important limitation that they did not discuss. In a study of this nature, patients with an interest or belief in meditation are more likely to consent to participate [a systematic error]; thus, the sample is likely to be biased toward meditation from the outset. This, coupled with the impossibility of patient blinding to treatment allocation, is likely to magnify the placebo response in the meditation group and correspondingly diminish the treatment response in the comparison group. The proper sampling procedure for studies in which patient blinding is impossible is to recruit only those subjects who are neutral in their attitudes. . . ."

(2) In their research on people, Herbert et al. (2006) discovered a correlation between a certain genetic variant (a qualitative state of X) and obesity (a quantitative state of Y). And as there is no conceivable construct Z causing both a genetic variant and obesity, and as obesity cannot cause a genetic variant, they concluded that the genetic variant causes obesity. Later, however, several other groups of researchers replicated the research, and they published comments on their findings in the 12 January 2007 issue of *Science* (p. 187). Two of the groups found no correlation, suggesting the genetic variant does not cause obesity. The other found a correlation of the genetic variant with being underweight, suggesting the genetic variant causes skinniness. Herbert et al. responded in the same issue of *Science* (p. 187) saying: "Identification of genetic variants affecting complex traits such as obesity is confounded by many types of bias [systematic error]. . . ." Clearly, some or all of the above researches are wrong. They are wrong because some or all of them failed to follow a best research practice: For experiments aimed at determining whether or not X causes Y, be sure to minimize the systematic error and the random error in the main constructs and the incidental constructs.

5.4 How the indirect method has a good deal going for it

I want to rebut opinions I have occasionally seen or heard that in effect say, "If you can't use the direct method, forget it." Here are three circumstances where application of the indirect method shines:

(1) The indirect method is ethical. As can happen in research with human subjects, ethics may stop us from applying the direct method. Imagine taking thousands of young adults, as alike as possible in lifestyles and genetics, and making half of them smoke a pack of cigarettes a day for 30 years. Prohibit the other half from smoking, and after 30 years count the number of cancer, stroke, and heart attack cases in the two groups. Yet it is ethical for people to decide for themselves to smoke or not, and researchers are ethically free to study them with the indirect method.

(2) The indirect method can be used in real settings. I have before me a research article titled "Smoke and pollution aerosol effect on cloud cover" (Kaufman and Koren 2006). It reports on a study that was carried out in nature, with the indirect method. Such a study cannot be done in the artificial setting of a laboratory, with the direct method. And as for research on the behavior of people or animals, they may behave artificially in artificial settings. It's often better to have them behave naturally in natural settings. And so it's often better to study natural cause-effect with the indirect method than to study artificial cause-effect with the direct method.

(3) The indirect method is often suitable for research into cause-effect in history. Take evolution. It was principally with observational studies that Darwin was able to retroduce his cause-effect theory of natural selection. As Thomson (1997) observes: "The primary test of responsibility [i.e., of cause-effect] is, of course, experiment. But what is a scientist to do when the matter under investigation is something unique and historical, no longer capable of experimental manipulation. . . . ? The principal tools available are comparative and analytical: the study of large numbers of analogous events, establishing a class of phenomena in which there are consistent associations of subphenomena. And, almost always in science, the key is to find things that can be measured. If A is always present in company with B and is never present without B, and if there is a strict proportionality of scale between the expressions of A and B, then we are a long way toward finding responsibility even though the methods used are purely observational."

It is generally lost sight of that application of the indirect method in the soft sciences – psychology, sociology, medicine, economics, ecology, biology, geology, and more – can be as convincing as applications of the direct method in the hard sciences – physics and chemistry. An image of how the soft sciences discover cause-effect comes from the way I have seen illustrators depict Gulliver tied down by the Lilliputians – a great many thin ropes irregularly placed, though highly effective. The image of how the hard sciences establish cause-effect truths is Gulliver tied down with one thick rope. What matters with either is that Gulliver is convincingly tied down.

Furthermore, the soft sciences are the only way we have to gain knowledge for solving most of society's important problems. Does exposure to lawn herbicides raise the incidence of lymphomas? Are the highly-strung more susceptible to strokes? Does being bullied in youth inhibit adult personality? What sorts of commercial developments endanger the lynx? Do children raised with the fine arts acquire empathetic values? Is kindness to people promoted by kindness to animals? The soft sciences can supply reliable answers to important practical questions of cause and effect – reliable enough to act on and improve the world.

5.5 Pseudoscientific claims of cause and effect

Every kind of research – conceiving hypotheses, testing hypotheses with the H-D method, testing for cause-effect, testing for whether or not chance produced an interesting pattern of facts, and deducing knowledge – is in danger for being pseudoscientifically fabricated. In this book we only have space to look at a slice of all the kinds of pseudoscience: chiefly pseudoscientific claims of cause-effect research.

Pseudoscience is half-baked science. Part is done with proper scientific protocol, part isn't. That's how it tricks people. They tend to shut their eyes to the improper part, letting the proper part rule their impression. The pseudoscientist does one or several things improperly, e.g., uses ad hoc constructs instead of standard constructs, or a non-random sample instead of a random sample, or a smaller sample than required, or doesn't minimize systematic error, or draws conclusions without consulting a baseline(s), and so forth.

Three sorts of people practice pseudoscience. (1) Wishful thinkers. Wouldn't it be nice if we knew that this herb cures cancer? A pseudoscience project can be made to show that it does. (2) Lax researchers. Those who don't strictly follow the best research practices – the proper methodologies and protocols – may at times unwittingly do pseudoscience. (3) Certain advocacy groups and political groups. To pull out an old term: snake oil salesmen.

Wishful thinkers. Here is one example of thousands: U.S. police departments have long wished for an accurate way of deciding whether suspects are telling the truth. For them, wish fulfillment was in the wings: a kind of linguistic polygraph was invented, a technique called statement analysis. Police trained in it ask a suspect questions, record and type out the answers, and apply statement analysis to determine the manner and the number of times supposedly incriminating words and phrases are used. Yet listen to Shearer (1999), a critic of statement analysis for its pseudoscientific underpinnings, including its lack of a factually determined baseline of effectiveness: "The validity and reliability of statement analysis has not been established. Most of what we know is based on theory at best and erroneous speculation at worst. It is not clear in statement analysis literature what is 'scientific' about the procedure. For example, we do not know the false positive or false negative rates for statement analysis – when investigators disbelieve a true statement or believe a false statement, respectively."

Lax researchers. Recently I was at a research seminar given by a master's degree candidate. He had invented a method for teaching children about nature. He designed an experiment to test the method in the local school system, and his advisor approved the design. After presenting the results of the test, he concluded that the new method was effective. Then his troubles started. A professor asked if he had used a control group. He explained he had, as follows: At the start of the project, students' knowledge of nature was measured with a test. After they were taught with the new method, they were retested. It was found they had learned. "That just shows," the professor said, "that your method does better than no method. I'm asking how the new method performs in a head-on comparison with the old method?" That part of the study, the student said, hadn't been done. And that is one form of pseudoscience – making a claim of the discovery of an effective method while failing to demonstrate it is better.

Another thing, every experiment in the world for which the sample size isn't properly set runs an unknown chance of generating false conclusions, and so chances being pseudoscientific. Take a cause-effect experiment that relies on regression analysis to uncover whether or not X and Y are correlated. If the researcher fails to set the sample size (with the help and guidance of a statistician, as discussed in chapter 2), the sample size may be less than what is proper. With that, the research will be overly prone to (1) giving a false positive result, suggesting that a statistically significant correlation between X and Y exists though actually it doesn't, or to (2) giving a false negative result, suggesting that no significant correlation exists though actually it does.

Cheating in science is wrong, but only a tiny fraction of researchers cheat. Were all cheating stopped, the overall quality of knowledge would improve just a little. Yet if somehow no scientist would ever settle for a smaller than proper sample size – or deviate in any other way from the best research practices – there would be great overall improvement in the quality of knowledge. (This is why I choose to teach best research practices instead of research ethics. I go where there is more to be gained.)

Certain advocacy groups and political groups.

Whenever advocacy or political groups practice pseudoscience, it is called junk science. Wanting scientific backing for their point of view, they loosely piece together parts of researches to arrive at an unproven claim, which they advertise as tried and true.

Let me give an example from my personal experience. It involves Orie, my cat. I was at the veterinarian clinic. Vet: "You should bring Orie in to have his teeth cleaned. He's got plaque built up." Me: "What's the problem? I don't like him to be put out [have anesthetics] at his age [15]." Vet: "It's very safe. If we don't remove the plaque he may develop periodontal disease. It can affect the heart, liver, and kidneys." Me: "Could it kill him?" Vet: "It could."

I told her I would let her know, and left the clinic thinking that Orie was under a death sentence unless I had his teeth cleaned. I assumed there was convincing veterinary research with the direct method, where a group of cats with heavy plaque, and another group with their plaque removed, were followed over time, and appreciably more cats in the first group died. I decided to see. I looked for the articles. I found none. Apparently no direct studies had been done. I did however find a PubMed research article under the byline of a certain veterinary dental clinic, with this statement: "The annual National Pet Dental Health Month program has resulted in a tremendous increase in public awareness. Veterinarians must carry this further in their everyday practices, convincing our clients of the need for preventive dental care [to fill our purses?]."

Besides, a journal article I found on gum disease in people noted: "Research shows that up to 30% of the population may be genetically susceptible to gum diseases. Despite aggressive oral care habits, these people may be six times more likely to develop periodontal disease." Given this holds roughly for cats, 30% or so are at risk for getting gum disease, plaque or no plaque. And while finding no research, direct or indirect, that showed the percentage of people, cats, or dogs with periodontal disease that died because of it, I did find research showing that anesthetic complications occurred in 12.0% of dogs and 10.5% of cats, and 0.4% of the dogs and the same percent of cats died from it. I called the vet. Confronted, she changed her tune. "I'm not surprised" were her exact words to my saying there's no conclusive research linking plaque to death. She had parroted the talk pushed by the American Veterinary Medical Association, never bothering to substantiate it.

Orie is now 20, with plaque. Naturally there's no way of telling if he is just a lucky exception. So I am not claiming that having vets remove plaque every year isn't a good precaution for some dogs and cats. I am claiming that reliable cause-effect research does not exist, at this time of writing, to know that the danger of plaque exceeds the danger of anesthetic complications. In plain words, the veterinary association's view rests on junk science.

We have to remember that claims of knowledge are nothing by themselves. We have to find out whether they were reached by scientifically sound research practices or by pseudoscientifically fallacious research practices.

5.6 Reciprocal cause-effect relations

This chapter's concern is the most common case of X causing Y. Yet reciprocal cause-effect relations deserve mention. They are when X causes a change in Y, and in response Y causes a change in X, and so on. George Orwell once gave this example: You feel yourself to be a failure, and this causes you to drink, and drinking causes you to feel more a failure, and this causes you to drink more, and so on, a spiral into drunken failure. Reciprocal cause-effect relations are the subject of much of the literature of systems analysis, ecosystems, and social systems, where effects feed back to influence causes. An easy to follow introduction is Dietrich Dörner's book, *The Logic of Failure.*

Even when X and Y do reciprocate, there is usually much to be gained in understanding the initial "X causes Y link," without going beyond that into the reciprocations. For example, recently I read a report of research titled: "Dirty Mouth Causes Pancreatic Cancer." Of course, perhaps more involved research would show that X and Y reciprocate, with a dirty mouth causing pancreatic cancer, and pancreatic cancer causing a more dirty mouth, and the more dirty mouth causing the pancreatic cancer to get a stronger hold, and so forth. But the important practical part of the research is the first step, that X causes Y.

6

How to decide if chance can easily produce that interesting fact

Imagine it's a minute after your research seminar. You talked about an interesting fact you observed. You proposed a research hypothesis, H, to explain it. Ordinarily you'd soon be outside, pleased it's gone well. Except someone in the audience says, "While you were presenting, I did some calculations that show that chance – blind fate, dumb luck – can easily account for the fact you observed." Now you're sweating. The person tested and corroborated the hypothesis of chance, H_c, raising doubt about your research hypothesis, H.

Or, imagine you test a research hypothesis, H, with the H-D method. You deduce a test prediction, P, and observe the actual fact, F. It turns out that P is true, corroborating H. You write the test up in a journal article. But just weeks after publication, the editor calls and informs you that a reader took the fact, F, and proved that chance could have easily produced it. Although you are invited to submit a rebuttal, the hole is too deep to get out of.

Now for some good news. Provided the hypothesis of chance can be tested (it can't always be), one of two methods will do it (which one depends on the nature of the fact, F). The names of the methods are less well established than the methods themselves. One we will call the random process method; the other, the base-rate method. The fact, F, they work with can be from experimental research, or from observational research, typically one-time events (Connor and Simberloff 1986, Romesburg 1989, Romesburg 1985).

Below, section 6.1 covers the random process method; section 6.2, the base-rate method. Section 6.3 explains and illustrates the kind of research where the hypothesis of chance cannot be tested. Finally, section 6.4 shows how testing the hypothesis of chance is a way of ending pseudoscientific twaddle about supernatural causes.

Concerning the definitions of several words that are basic to the methods, by "fact" we mean a single fact or a collection of facts. Accordingly, each fact in a series of facts is a fact, and the series of facts is itself a fact. And by an "event" we mean a "fact." For instance, the event of Vesuvius erupting is the fact of Vesuvius erupting. Finally, by "data" we mean a set of facts, e.g., a list of dates when Vesuvius erupted.

6.1 The random process method of testing the hypothesis of chance

Both methods – the random process method and the base-rate method – require a way of generating what we will call "a sample space." A sample space is a list of facts, any of which could have possibly occurred if the hypothesis of chance is true. For a sample space consisting of, say, 50 possible facts, each has a probability of 0.02 of occurring (1 divided by 50).

With the random process method, there are two approaches to generating a sample space. One we will call the permutation approach; the other, the random number approach. They are as follows.

The permutation approach.

Let's first review what a permutation is. Suppose we have three items: a dog and two people: Spot, Sue, and Phil. There are 3 x 2 x 1 = 6 permutations of the three: (1) Spot, Sue, Phil, (2) Spot, Phil, Sue, (3) Sue, Spot, Phil, (4) Sue, Phil, Spot, (5) Phil, Spot, Sue, (6) Phil, Sue, Spot.

With five items, there are 5 x 4 x 3 x 2 x 1 = 120 permutations; with seven items, 7 x 6 x 5 x 4 x 3 x 2 x 1 = 5,040 permutations; with 12 items, 12 x 11 x 10 x 9 x 8 x 7 x 6 x 5 x 4 x 3 x 2 x 1 = 479,001,600 permutations; with 30 items – Hold on to your hat! – there are more permutations than grains of sand in all the world's beaches.

To use the permutation approach, the observed fact must be composed of elemental facts. Each permutation of the elemental facts is a point in the sample space. In cases where all possible permutations is a mind-boggling number, we instead generate a smaller, representative set of permutations. In still other cases, it is unnecessary to generate any permutations; we simply imagine that we have (as we'll illustrate shortly). The research examples below show how the permutation approach goes:

Res. Ex. 6.1: Will your monkey be a gold mine?

Suppose that Adora is your pet monkey. One day in the rumpus room she finds four children's blocks. One of them has the letter A on its six sides, one has E on its sides, one has M, and one has D. She strings them together, spelling the word MADE. That's the observed fact; you are excited about it. But is your excitement justified: Was it a case of choice or chance? The two competing hypotheses are:

H: Adora relied on intelligence to spell MADE.

H_c: Adora accidently spelled MADE.

You decide to test H_c, the hypothesis of chance. The test has three steps:

(1) Note that the observed fact – MADE – is composed of four elemental facts: one is the M, one is the A, one is the D, and one is the E.

(2) List all permutations of the elemental facts. There are 4 x 3 x 2 x 1 = 24, and they make up the sample space:

MADE	ADEM	DEMA	EMAD
MAED	ADME	DEAM	EMDA
MDAE	AEDM	DMEA	EAMD
MDEA	AEMD	DMAE	EADM
MEAD	AMDE	DAEM	EDMA
MEDA	AMED	DAME	EDAM

If H_c is true, each of the 24 is equally likely to have occurred.

(3) Count the number of the 24 permutations that, had they occurred, would have excited you – excited you because they would make you suspect that Adora is intelligent. There are four: MADE (the one that did excite you), MEAD (an alcoholic drink), DAME (a woman of rank), and EDAM (a type of Dutch cheese). Thus the probability that chance could have produced a word that would have excited you is $4/24 = 0.167$ (16.7 percent).

Had the probability been at or less than the statistical standard of 0.05 (5 percent), you would feel surer that Adora is intelligent. But four in 24 fails to overthrow H_c, leaving you thinking, "She may have lucked out."

Suppose instead she found nine blocks with the nine letters E, I, D, H, R, N, B, G, and L on them, and gone right to work and come up with LINDBERGH. Start thinking $$$$$. There are 362,880 ways of stringing them together, too many permutations to list, and only one of them spells a word (I'm guessing). Compared to four blocks, nine blocks make a harder challenge for chance to struggle against, making it harder for Adora to luck out.

Comment: Several nonparametric statistical methods are based on permuting elemental facts. Chief among them are what statisticians call bootstrap tests and randomization tests. They are for testing statistical hypotheses about a population's parameters, such as the mean, μ. The tests require the facts to be a random sample of the population. Yet testing the hypothesis of chance with the permutation approach, for all its operational resemblance to nonparametric tests, has a different aim than estimating a population's parameters, nor does it require the observed fact to be from a random sample. The example of Adora illustrates this different aim. The test results don't generalize to all monkeys; they apply only to a particular monkey, Adora. Still, that can be scientifically valuable.

Res. Ex. 6.2: What's up with the foxes?

In the early days of wildlife research, animal ecologists made graphs of the estimated population sizes of Canadian arctic foxes over long series of years. A bumpy succession, the number of foxes peaked, dropped for one or more years, made a peak again, and so forth.

When the ecologists computed the average number of years from one peak to the next peak, they discovered it was about three years. Their surprise grew as they found the same observed fact – an average of three years between peaks – in population size data of other arctic predators, e.g., wolf and lynx. To account for the observed fact, they retroduced the following research hypothesis:

H: Unknown environmental and/or genetic factors produce an average peak-to-peak interval of nearly three years.

After fruitlessly beating their minds for what the factors could be, one animal ecologist decided to test the hypothesis of chance:

H_c: Chance produces an average peak-to-peak interval of nearly three years.

Let's sketch how the test goes. We'll assume a 100-year run of population size data for foxes. Ideally we would take the 100 values and list all permutations of them. That is an astronomical number. So instead, we program a computer to randomly generate 10,000 permutations, a smaller set whose properties will acceptably reflect those of the full set. For each of the 10,000 permutations, each 100 years in length, we compute the average peak-to-peak interval. When we do this, we see that in practically all of the permutations the average peak-to-peak interval is about three years. By this, H_c is corroborated, and so we conclude that H is probably false. (Mathematically, Kac (1983) put the conclusion beyond dispute when he proved that for an infinitely long list of random numbers, the distance between peaks is exactly three.)

Rather than actually permuting the elemental facts that make up the observed fact, sometimes it suffices to imagine what would happen if we did the permuting. Here's an illustration:

Res. Ex. 6.3: Were Africa and South America once one?

As every school child notices, the fact is that the west coastline of Africa and the east coastline of South America are a pretty close fit. Aha! A hypothesis:

H: Africa and South America were once joined.

History records that Abraham Ortelius was the first known person to utter the Aha! Writing in 1596, he proposed that the two had been joined; later, others independently suggested the same (Romm 1994).

Why is H very convincing? It's because the hypothesis of chance is very unconvincing:

H_c: That the two coastlines fit closely is just happenstance.

Why do we believe H_c is false? My opinion is that we unconsciously imagine steps like the following. Imagine the outline of the west coastline of Africa is cut into, say, ten pieces. Imagine the ten are permuted in all possible ways. Imagine that for each permutated order, the pieces are joined, giving a permuted outline of a coast. The number of permuted outlines is 3,628,800. Likewise, imagine the same is done for the outline of the east coastline of South America.

Imagine now that each of the 3,628,800 African outlines is compared with each of the 3,628,800 South American outlines. Imagine we record how many of the comparisons are as closely hand-in-glove fitting as the actual observed fit is, as we see it on the world map. Practically none will fit as well as the actual coastlines do. Hence, it's a safe bet that the close fit of the actual coastlines is not happenstance. Hence, it's a safe bet that they were once joined. (Interestingly, the word "happenstance" is an accidently altered form of the older word "happenchance," short for "a chance happening.")

Res. Ex. 6.4: Why do students sit in clusters?

On the first day of my college classes, I see the students file into the rooms and choose seats. The interesting fact is that the males tend to sit together in separate clusters of two to six or more, and the females do the same. Once, observing this, I proposed a research hypothesis:

H: Social processes explain the sexual clusters.

The social processes I had in mind happen outside the classroom. Namely, females tend to walk and chat together from one class to the next, and males tend to do the same. Each group comes into the classroom and plops down, producing clusters of females and males.

At the same time, maybe social processes have nothing to do with it. Spatial clusters can form by chance. I had better, I thought, test the hypothesis of chance:

H_c: Chance explains the sexual clusters.

Previously I had written a computer program, ZORRO, that tests the hypothesis that spatial clusters form by chance (Romesburg 1989). I decided to run it. The room's seats were on a rectangular grid, and on the first day of a class I recorded the sex of the student occupying each seat. I fed these data into ZORRO, and it computed a numerical index of the degree of clustering of males and females. Next, ZORRO randomly permuted the observed arrangement of students in the room, reassigning each of the students to an originally occupied seat. In all, ZORRO did 10,000 permutations, and for each it computed the degree of clustering of males and females. In less than 5 percent of the 10,000 permutations was the degree of clustering as pronounced as, or more pronounced than, the observed degree. This cast doubt on the possibility that the clustering was due to chance, thereby providing assurance that H was true.

A related point: Among what ZORRO will do is test the hypothesis of chance concerning spatial clusters of species of trees growing on land. It will also test the hypothesis of chance concerning temporal clusters of species of birds returning to sites after overwintering in distant locations. In a similar vein, a computer program named CLASSTEST (Romesburg 1985) is for testing the hypothesis of chance in cases where the observed fact is made up of multivariate data.

The random number approach. For the random number approach to testing the hypothesis of chance, we do not permute the elemental facts that comprise the observed fact. Rather, we do a probabilistic simulation with the elemental facts, using random numbers to randomly shift the elemental facts around in space or in time. In some cases we can dispense with actually doing this, and instead just imagine that we did it. The next seven research examples demonstrate the random number approach.

Res. Ex. 6.5: Why are the planets' orbits in nearly the same plane?

Let's take our computer along as we travel back to Daniel Bernoulli's time, the early 18th century, when there were six known planets. An interesting fact is that their orbits lie in nearly the same plane. We venture a reason:

H: As a result of astrophysical processes, the planets' orbits lie in about the same plane.
Before getting down to working out what the processes could be, we should test the hypothesis of chance:

H_c: As a result of chance, the planets' orbits lie in about the same plane.

To test H_c, we have our computer carry out a series of steps. Step 1: Compute the average separation, in degrees, of the six observed orbital planes. Step 2: Use random numbers to randomly orient the six planes in space, and for each random orientation compute the average separation of the planes. Step 3: Repeat step 2 many times, say 10,000. Step 4: Compute the proportion of the 10,000 times for which the average separation of the orbital planes is as small as, or smaller than, the observed separation. In case the proportion is less than 0.05 (5 percent of the 10,000 times), decide that H_c is false, i.e., that chance is unlikely to have oriented the planes.

Hence, decide that H is true, i.e., that unspecified astrophysical processes most likely oriented the planes. Conversely, in case the proportion is larger than 0.05, decide that we cannot reasonably dismiss the idea that chance oriented the planes.

An alternative to using random numbers is to do the test mathematically, applying probability theory to calculate an exact probability of chance. Listen to Box (1978) tell how Daniel Bernoulli did just that: "Before considering the essay question set by the French Academy as to why the planetary orbits were nearly but not exactly in the same plane, [Bernoulli] questioned whether the orbits might not reasonably be considered to be a random arrangement. To test this hypothesis, that the planetary orbits owe the degree of their coplanarity to chance alone, he calculated the dispersion of the points of intersection with a unit sphere of the poles of the planetary orbits, and thence the probability that points clustered as closely as or more closely than these were on the surface of the sphere would occur by chance. He concluded that the probability was too low to be considered plausible and that the planetary orbits could not be considered as random. Having demonstrated the reality of the phenomenon of nonrandomness by the test of significance, he felt justified in considering the reasons for the degree of coplanarity observed."

A further point: For this research, the permutation approach will not work. Look what happens when we try to apply it. We permute the order of the six planets from the sun, giving 6 x 5 x 4 x 3 x 2 x 1 = 720 permutations. Each permutation assigns each planet to an orbital plane, for instance Earth to Mars' orbital plane, Venus to Earth's, Saturn to Jupiter's, Mercury to its own orbital plane, and so forth. The trouble is, the average separation in orbital planes is the same for every permutation. This won't do for testing the hypothesis of chance.

Comment: The above planetary research well typifies the value of testing the hypothesis of chance. On the one hand, deciding chance oriented the orbital planes would save needless work in trying to support the research hypothesis, H. On the other, deciding chance didn't orient the orbital planes would let us address our work to trying to determine the astrophysical processes that did.

Comment: More than a century before Daniel Bernoulli, Isaac Newton examined the fact that the six known planets of the time moved in the same direction around the sun. In his book, *Opticks*, published in 1704, he concluded, "Blind fate could never make all the planets move one and the same way in orbits concentrick." Under the hypothesis of chance, the probability of the six moving in the same direction is $(0.5)(0.5)(0.5)(0.5)(0.5) = 0.03125$, a figure small enough to convince most people that chance played no hand in arranging the direction of travel.

Comment: Type into the Google search window the words "archeology" and "random orientation" and the search will return various applications of the random number approach. Archeologists dig up tools, bone fragments, graves, and so forth, and always the first question to resolve is, are they randomly or deliberately oriented?

Res. Ex. 6.6: How did we learn that atoms have a nucleus?

The discovery of the nucleus is, in its way, quite a striking story. Two physicists working for Ernest Rutherford in 1909, Hans Wilhelm Geiger and Ernest Marsden, aimed a beam of alpha particles at a flat piece of gold foil. Most of the alpha particles passed through, but about one in every 20,000 did a U-turn. To explain this fact, Rutherford retroduced this hypothesis:

H: The weight of an atom is concentrated at its center, its nucleus. Surrounding the nucleus is much empty space, where the atom's electrons orbit, giving most of the alpha particles free passage through the foil. Those that reverse direction inside the foil and emerge from its front surface do so because they hit a nucleus and bounce back.

But H had a contender, the hypothesis of chance:

H_c: The alpha particles that emerge from the foil's front surface randomly undergo small-angle glancing deflections with nuclei and electrons in the foil, the cumulation of which turns their direction around.

To test H_c, Rutherford computed the probability that small-angle deflections within the foil could produce a U-turn. He got this probability: 3×10^{-2174}. How small is that? "Utterly negligible," wrote Weinberg (2003), explaining that: "Even if all the material of the universe consisted of alpha particles and each alpha particle was fired billions of times per second through this gold foil, the chance that such an improbable event would have occurred even once in the history of the universe would be utterly negligible."

Thus this disproof of the hypothesis of chance much strengthens belief that H is true.

Res. Ex. 6.7: Can large earthquakes waken distant volcanoes?

Linde and Sacks (1998) examined 400-year global records of dates when powerful earthquakes occurred, and examined records of dates when volcanoes erupted within 750 kilometers of each powerful earthquake's epicenter. They saw a pattern: Of 204 earthquakes of magnitude 8.0 or greater, 8 of them were accompanied on the same day with a total of 11 volcanoes beginning to erupt. Besides, only 5 volcanoes within 750 km erupted on any of the 1,000 days before or after the 8 earthquakes. To explain this fact, Linde and Sacks proposed:

H: Volcanoes poised for a large eruption may be pushed over the brink by powerful earthquakes.

Linde and Sacks' next concern was to decide whether or not the hypothesis of chance was true:

H_c: The observed earthquake-eruption pattern is indistinguishable from randomly occurring earthquake-eruption patterns.

Linde and Sacks tested H_c with a probabilistic simulation of earthquake-eruption patterns. They programmed a computer to randomly select dates before and after each of the earthquakes, and to record the number of eruptions on these dates. Repeating this in 100,000 trials, not once did chance produce a day so dense in eruptions as in the observed data. This put H_c (dare we say it?) on very shaky ground, and hence strengthened belief in the research hypothesis, H.

Comment: With probabilistic simulations to test H_c, how many trials are necessary? 100? 1,000? 10,000? 100,000? The answer is that 10,000 suffice (Romesburg 1985). For the above earthquake-eruption association study, 10,000 would have been enough.

Res. Ex. 6.8: Is the alignment of the ancient tomb by design?

A tomb uncovered at Newgrange, Ireland, dating from 3,150 B.C., set archeologists buzzing about a curious fact. On its east-facing side, a slit in the stone would have allowed a beam of light to enter at sunrise on the winter solstice, according to the solar alignment at the time, and shine down a narrow passage to illuminate the floor of the main chamber, 18 meters away. To Ray (1989) this suggested a research hypothesis:

H: The tomb's alignment to the sun's was deliberate.

This hypothesis makes no testable predictions. Nor can we slip back five thousand years and ask around. But it has an opposite hypothesis that is testable:

H_c: The tomb's alignment to the sun was accidental.

To test H_c, Ray let chance orient the tomb, and only in 0.83 percent of the simulated trials could sunlight have reached the floor of the main chamber. Accidental alignment, it follows, is nearly out of the question. Hence, deliberate alignment is easy to accept.

Res. Ex. 6.9: Are those French churches equidistance on purpose?

In his book, *The Holy Place*, Henry Lincoln mentions the interesting fact that for all of the more than one dozen pairs of churches in a seven square-mile area of France, the two churches in each pair are the same distance apart. Lincoln goes on to propose that this is deliberate. However, a reviewer of the book guessed it might well be accidental.

Lincoln defended his proposal of deliberate spacing in a letter (Lincoln 1991), challenging the reviewer to use dividers to measure the pairwise distances on a map of the area. Lincoln ended his letter with this sentence: "If, having made this simple test – and without involving himself in any further complexities of the geometry – he [the reviewer] still insists that this repetition of a fixed measure (even on a map) is the result of wishful thinking on my part and is nothing more than pure coincidence, then I can only suspect that he is of that confraternity of 'rationalists' who would also insist that their mother's presence at their birth was equally coincidental."

Comment: Assume, for a moment, that we observe a fact for which there's no imaginable physical explanation, leaving no possibility of a research hypothesis, while the test of the hypothesis of chance strongly suggests that chance does not account for the observed fact. That would put us in a jam, and pressure us to believe that the result of testing the hypothesis of chance is a statistical fluke. I am thinking of an instance of this in astronomy, which Schilling (2006) has written about: When astronomers discovered the first gamma ray burst (GRB) in the distant universe, they observed that its light passed through a galaxy in reaching us. When they discovered the second GRB, its light, too, passed through a galaxy in reaching us. More and more, astronomers became surprised as each successive discovery of a GRB (at this writing, 14 of them) originated at a distant place where the light from it passed through at least one galaxy in reaching us. Hard as it is to swallow that this is because of chance, it is harder to swallow that there is a physical reason for it. Says astronomer Ken Lanzetta: "If I had to bet, I would say that this is a one-in-10,000 statistical fluke that happens every now and then. It will probably go away when more observations become available. We'll have to wait and see." He adds that if the next two dozen or so GRBs are all lined up with galaxies, then "something very strange must be going on."

Res. Ex. 6.10: Do dowsers have genuine skill?

Here's a research hypothesis that many country folk, farmers, and ranchers accept as true without scientific proof:

H: Professional dowsers have actual skill for finding water.

Against this view, Enright (1999) proposed the hypothesis of chance:

H_c: Professional dowsers find water no more often than random guessing can.

To test H_c, Enright employed forty-three professional dowsers to participate in a dowsing experiment made along a 10-meter test line covered by a wooden floor. For each trial he hid, at a randomly determined point on the line and below the floor, a pipe containing running water. He had a professional magician eliminate any subtle clues to the water's location. Each dowser performed with no other dowser around. Double-blind procedures were followed; neither dowser nor accompanying observer knew where the water was. In all, the dowsers made 843 dowsing trials, each ending with the dowser saying where the water was located. From this he computed their success rate. Then he determined the success rate of randomly guessing the locations of the buried water. The dowsers fared no better than randomly guessing. This proves H_c, the hypothesis of chance, and disproves H, the hypothesis of skill. (See Letters-to-the-editor (1999) for comments on Enright's study.)

Res. Ex. 6.11: How did the cat find its way home?

A newspaper report (Associated Press 1995) tells of a couple that stopped their car at a rest area in Jackpot, Nevada (a place that thrives on chance), while driving home to Mountain Home, Idaho, 95 miles away as the crow flies, and 132 miles by road. Their cat jumped out and ran away. Thirty-five days later it trudged up to the couple's front door. From this fact, it is natural to retroduce the following hypothesis, H:

H: The cat drew upon an instinct to find its way home.

But before accepting this as true, we should test the hypothesis of chance, H_c:

H_c: Straying aimlessly, the cat accidently stumbled upon its home.

To test this, we could devise a probabilistic simulation to estimate the probability that a cat, making random decisions of when to change direction and how far to go before changing again, could accidently arrive at its home. But skip the program; common sense tells us there's practically no possibility. Thumbs down, therefore, on the hypothesis of chance, H_c. And with that, thumbs up on the instinct hypothesis, H. (Besides, there are other corroborating instances of wayward cats returning to distant homes.)

This finishes our discussion of the random process method of testing the hypothesis of chance. We have seen how the two approaches to the testing go: the permutation approach and the random number approach.

6.2 The base-rate method of testing the hypothesis of chance

With the base-rate method, we do nothing to the observed fact. It cannot be reduced to elemental facts. That prevents us from applying the permutation approach and the random number approach. Rather, we compare the observed fact to base-rate data, viz., to the empirical frequency distribution of facts of its kind. By this, the frequency distribution reflects the sample space.

Where do the base-rate data come from? Perhaps someone has compiled them. If not, perhaps we can compile them. Other times we may have a reliable intuitive sense of them, and can mentally sketch the frequency distribution well enough to reach a decision on the test of the hypothesis of chance.

In the four illustrative research examples below, a fact is observed and a hypothesis, H, is retroduced that explains the fact. Along with this, the hypothesis of chance, H_c, is proposed as a competing explanation of the fact, and the base-rate method is used to test H_c.

Res. Ex. 6.12: The World Trade Center disaster: chance or planned?

Almost at the minute we learned of it, why were we certain that the World Trade Center disaster was deliberately caused? It's because within us we carry base-rate data of the times between airplanes accidently hitting large buildings. My use of mine is typical. On the morning of September 11, 2001, the TV program I was watching was interrupted with pictures of smoke coming from the north tower of the World Trade Center, along with the news that an airplane had crashed into it. That was the observed fact.

As to base-rate data, I remembered an accidental crash of a military bomber into the Empire State Building in 1945. I remembered another accidental crash of a cargo plane into an apartment building in Amsterdam in the early 1990s. What's more, I could not recall any deliberate crashes. This was it, my base-rate data: decades passing between two accidental crashes into large buildings, and no deliberate crashes.

As I watched, a hypothesis, H, occurred to me that the crash into the north tower might be deliberate. Against this, another hypothesis, H_c, occurred that the crash was one of those rare accidents. I think I favored the accident hypothesis; it had historical standing. Then within ten minutes came the report that the south tower had been struck. That brought a new observed fact. One crash was within the norm of my base-rate data for accidental crashes. Two crashes in nearly the same place and at nearly the same time was far outside the norm. Therefore, these weren't accidents; they had to be deliberate.

Res. Ex. 6.13: How did Kimberly Bergalis get infected with HIV?

In 1993, Kimberly Bergalis' dentist, Dr. David Acer, who had AIDS, was tried and condemned in the U.S. national media for having infected her with HIV. This happened despite virtually no one knowing Bergalis's history of sexual contacts. The logic was guilt by association. The media's hypothesis was:

H: Dr. Acer transmitted HIV to Kimberly Bergalis.

The hypothesis of chance was:

H_c: In Kimberly Bergalis' life outside Dr. Acer's office, she accidentally became HIV-infected.

In a letter to *The Wall Street Journal* responding to a story the *Journal* had run previously, Bland (1993) argued that base-rate data support the hypothesis of chance. Bland noted that the incidence of HIV in the U.S. population was 0.4 percent. Going on, he said that Dr. Acer had about 2,000 patients, and six (including Bergalis) were known to have HIV, which is 0.3 percent, less than the national base-rate. Thus there is a good possibility that H_c is true, and a corresponding good possibility that H is false.

Res. Ex. 6.14: What about O. J. Simpson's shoes?

A partial imprint of the sole of a size 12 Bruno Magli shoe was found at the scene of the murders of Nicole Simpson and Ron Goldman. It was known that O. J. Simpson wore size 12 shoes, and pictures existed showing him in Bruno Magli shoes. This suggests the following hypothesis:

H: The partial imprint was made by O. J. Simpson.

Against this is the hypothesis of chance:

H_c: The partial imprint was made by someone else, such as a chance passerby.

To test H_c, prosecutors in Simpson's trial located sales records of size 12 Bruno Magli shoes (base-rate data). If none or few had been sold, that would go against the hypothesis of chance, in turn strengthening the hypothesis that the imprint was Simpson's. As it turned out, enough size 12 Bruno Magli shoes had been sold that the hypothesis of chance could not be dismissed.

Comment: Along the lines of identifying people by partial shoe imprints is identifying people by partial fingerprints. A person is accused of a crime because a partial fingerprint found at the scene matches his or hers. Yet what is the likelihood that it was made by another person? To answer this, we need to consult base-rates of the frequency at which people have the same partial fingerprints. What proportion of the population (or relevant subpopulations) has particular ridge characteristics? Faigman (2002) answers: "This is a question of base-rates. To be admissible [in a court of law], fingerprint identification need not be powerful enough to show identity, but the fact-finder should be given some idea whether one person in 5, or 100, or 1000, could have left the partial print."

Res. Ex. 6.15: Are plagiarists in the classroom?

And now I have a little surprise for you. You are going to play the role of a teacher. Your class is World History, and you ask your students to write a 1000-word term paper on the probable course of history had Franz Ferdinand not been assassinated. Later comes a disturbing fact: Grading the papers, you notice the same 32-word passage in two of them. Automatically, a hypothesis springs to mind:

H: The authors of the papers plagiarized.

You call the two in. Welcome to teaching!: They swear their papers are entirely original. One says, "History is filled with coincidences. This is one of them."

Having read this book, you always test the hypothesis of chance whenever feasible, as it is here:

H_c: By pure coincidence, the authors wrote identical 32-word passages.

To test this, you need base-rate data. You get it by estimating the frequency distribution that by pure coincidence the two papers could have X-number of words in a series be the same. You realize that with X as small as ten words, the likelihood is far less than one in a hundred. That tells you that the likelihood of having two 32-word passages the same is incomprehensibly small. So you are as sure as can be that H_c is false, and by that you know that H is true. It must be that either one student copied from the other, or both copied from a paper not among those you graded, perhaps one on the Internet.

6.3 Where the hypothesis of chance cannot be tested

In his book, *Mathematics and Plausible Reasoning: Volume II Patterns of Plausible Inference*, George Polya has a chapter titled, "Chance, the Ever-Present Rival Conjecture." A conjecture is a hypothesis. He could have called the chapter, "Chance, the Ever-Present Rival Hypothesis." Yet he is wrong about the "ever-present": "sometimes-present" is what it should be. In certain research, the observed facts are such that the hypothesis of chance cannot be tested. For an illustration, think back to Res. Ex. 3.11, called "Why is the night sky dark?" It lists several retroduced hypotheses that are able to account for the interesting fact that the night sky is dark. Yet we can't test the hypothesis of chance:

H_c: It is purely by chance that the night sky is dark.

We can't test it because there is no freedom to let chance have a go at it, sometimes producing a dark night sky, sometimes not. In particular:

(1) We can't use the random process method to test H_c. First off, the permutation approach will not work: the observed fact – the dark night sky – has no elemental facts. And without them, we have nothing to permute. Nor will the random number approach work: it's nonsense to think of randomly shifting the observed fact – the dark night sky – about. Shift it about what?

(2) We can't use the base-rate method because we can't get the frequency of occurrence of facts like the observed fact of the dark night sky. For that we would need access to other universes and note the degree of light in their night skies. Moreover, there may not be other universes.

All the same, whenever it is feasible to test the hypothesis of chance, do it.

6.4 How to debunk hypotheses that propose supernatural causes

One of the great faults of more than one billion people is their tendency to uncritically embrace the notion that miraculous events have supernatural causes. The next four subsections deal with debunking claims of supernatural causes. The first explains what miracles are. The three that follow the first cover three sequential steps to debunking claims that miracles were caused by supernatural agents. In brief:

Step 1: Try to show that natural laws explain the apparent miracle. If you succeed, you are finished, with no need to go to step 2.

Step 2: Use base-rate data, if available, to test the hypothesis of chance, determining whether or not chance explains the miracle. If you decide chance explains the miracle, you are finished, with no need to go to step 3.

Step 3: Apply Littlewood's Law of Miracles. It is always applicable, and it always results in disproving the existence of supernatural agents. Equivalently, that there is Littlewood's Law implies there are no supernatural agents.

What miracles are. Three properties define miracles. A **miracle** is an (1) improbable, (2) marvelous happening that is (3) seemingly unexplainable by the laws of nature. Put another way: An event happens that has never happened before in our experience, or at most very infrequently. We marvel over it, and, as far as we know, no branch of science can explain why it happened. It's a miracle.

I have two things to say about miracles. The first is that miracles come in good, bad, and dispassionate varieties:

A good miracle is the only kind of miracle that religions recognize. Think of Lourdes, where a miracle is the cure of someone's usually incurable disease.

A bad miracle is when, for example, in the space of 24 hours a woman's son dies in a car crash, her husband learns he has prostate cancer, and a hurricane levels her parents' home.

A dispassionate miracle is the kind of strange, marvelous event you read about in newspapers. A golfer, say, has never driven a golf ball more than 215 yards, and has never scored a hole in one. Presto, out of the blue, on a 230-yard hole the golfer makes a hole in one.

The second thing I have to say about miracles is that the event of a miracle may be a single event or a coincidence of single events. Here are examples:

A single event: A farmer's chicken lays 524 eggs in a year. It's a single event: it's about one kind of thing, egg laying. And it's improbable: 524 eggs in a year is abnormally far out in the slim tail-area of the frequency distribution of egg laying. And it's marvelous: "My word!," exclaims the farming community. And science seems to have no physical explanation: "Beats me," says the poultry science expert, when asked how the chicken could do it.

A coincidence of single events: You are at home reading *Sky & Telescope* magazine, and a meteor comes through the roof. "What a coincidence!" you think. You wouldn't be blamed for calling it a miracle. Reading *Sky & Telescope* is one kind of event. A meteor falling through your roof is an entirely different kind of event. Both are in the same area of thought, astronomy. As well, the coincidence is improbable: No one has ever seen or heard anything its equal. The coincidence is marvelous: Of course it is – you're being interviewed on the TV news about it, aren't you? The coincidence defies natural explanation: Nothing in science links reading and rocks falling through the roof.

Coincidences, when their component events are in unrelated areas of thought, tend not to be marvelous, i.e., not miracles. You wouldn't have said, "What a coincidence!" if you had been reading *Time* magazine when the meteor came, or had been reading *Sky & Telescope* when a piece from a passing airliner fell though the roof.

So much for what miracles are. The kind of miracle that interests us from here on out is the kind that some people claim is supernaturally caused. Mostly, these are good miracles, with loving supernatural agents. Yet some are bad miracles, with vengeful supernatural agents. (Dispassionate miracles won't interest us, although at times I have heard of rare, marvelous shots in golf being attributed to "the golfing gods.")

Step 1 for trying to debunk a claim of a supernatural cause. Step 1 is to try to show that the apparent miracle is explainable by the laws of nature. If you succeed, conclude that the apparent miracle is just an improbable, marvelous event, entirely accounted for by science.

Here's an illustration: Edgar is present as a friend is being buried, and he sees several hundred fussing crows sweep into the cemetery and land in the trees. The event is a marvel; it's improbable; he has no explanation for it. That makes it a miracle to him: a bad, frightening one that he takes as an ominous sign sent by a supernatural agent. At the same time, standing nearby is Susan, a biologist. It's no miracle to her. She knows why the crows are there. She watched them chase a hawk into the cemetery. To her, it's simply two unrelated coinciding events, each with a natural cause. The funeral goers are attending to their business, burying a friend. The crows are attending to their business, mobbing a hawk that was threatening their nesting babies. The two explainable businesses just happen to be occurring at the same time and place.

Step 2 for trying to debunk a claim of a supernatural cause. Whenever step 1 fails to debunk a claim of a supernatural cause, go to step 2: test the hypothesis of chance. In all of the times I have seen step 2 taken, the test has been made with base-rate data (although, in principle, the random process method could in certain cases be used). Naturally, a condition for step 2 is that base-rate data are available. The next four research examples demonstrate step 2.

Res. Ex. 6.16: Lincoln and Kennedy: an occult connection?

Martin (1998) reports a set of related coincidences about the lives of Abraham Lincoln and John Kennedy. All are improbable. No one of them is particularly marvelous, yet together they seem to border on the miraculous. No laws of nature can explain them.

To begin with, both Lincoln and Kennedy have seven letters in their last names. Both championed civil rights. Both were assassinated, both with bullets to the head, both on a Friday, and both in front of their wives. Lincoln's secretary was named Kennedy, and Kennedy's was named Lincoln. The assassins, John Wilkes Booth and Lee Harvey Oswald, went by three names. Both have a total of 15 letters in their names, and both were from the South. Booth shot Lincoln in a theater and fled to a warehouse; Oswald shot Kennedy from a warehouse and fled to a theater. Andrew Johnson and Lyndon Johnson, who assumed the Presidencies, were both from the South, both Democrats, both with same last name, and both had six letters in their first names. Nor is this all. Lincoln and Kennedy were born in 1808 and 1908, 100 years apart. They were elected to Congress in 1846 and 1946, 100 years apart. They were elected to the Presidency in 1860 and 1960, 100 years apart. Andrew Johnson was born in 1808, and Lyndon Johnson was born in 1908, 100 years apart. Booth was born in 1839, and Oswald was born in 1939, 100 years apart. To account for these facts, one hypothesis that can be retroduced is:

H: The facts were caused by a supernatural agent.

A counter hypothesis is the hypothesis of chance:

H_c: The facts are purely coincidental.

To test H_c, we consult base-rate data on the expected number of coincidental similarities between any two people. We see that many are expected. It's well verified that combing through the personal attributes and life histories of any two people inevitably turns up many coincidental similarities. It might be that both have degrees in psychology, both are Quakers, both have an aunt that went to Bryn Mawr, both are fourth grade teachers, and so on. Or put another way, what would be extremely improbable is combing through any two people's personal attributes and life histories and finding few coincidental similarities. Therefore, concerning the coincidences surrounding Lincoln and Kennedy, it's dead certain that H_c is true. So rest assured, H is false.

Res. Ex. 6.17: King Tut: curse or coincidence?

In case you raised an eyebrow at my saying miracles don't have to be good, here comes a story of what many people took as a bad miracle caused by a vengeful supernatural agent. It begins with the expedition that discovered King Tut's tomb, which was the brainchild of George Herbert, better known as Lord Carnarvon. As the story built, it fired the public's fascination with the occult. P. W. Wilson (1930) summarizes the facts:

"Near Newbury there is a hill, notable in the landscape, where in the days of good Queen Bess they kindled a beacon to warn old England of the Spanish Armada. On that beacon Carnarvon himself had lit a patriotic blaze to celebrate the Armistice. Little did he think that it was to be, in a sense, his funeral pyre, and that, alone in his glory, he was to be buried on that Beacon Hill as the discoverer of the tomb of King Tutankhamen.

"Lord Carnarvon had a half brother, a member of Parliament, also quite a wit, Colonel Aubrey Herbert. Naturally he visited the tomb and, just another coincidence, he fell ill and died.

"There was a millionaire, with a name known throughout the world, George Jay Gould. He also did a little sight-seeing around the tomb and, a third coincidence, he died. Another man of means, Woolf Joel, yachting on the Nile, gratified his curiosity, examined the tomb, and become the fourth coincidence. The leading specialist, Sir Archibald Douglas Reid, was selected to apply x-rays to Tutankhamen's bones. But he died. The director of the Egyptian Section at the Louvre, Professor Georges Benedite, was enthusiastic over the tomb, and he died. Professor Lafleur of McGill University happened to be at Luxor and visited the tomb. At Luxor, he died. H. G. Evelyn-White, scholar and Egyptologist, left a letter with the words, "I knew there was a curse on me," and he committed suicide. M. Pasanova, who had helped to excavate, was another victim. The toll had reached nine, yet even so it was far from complete.

"The leader of the expedition was Howard Carter. His secretary had been the Hon. Richard Bethell, a man of forty-six years and therefore in the prime of his life, an Etonian, also of Oxford and the Guards, eldest son and heir to Lord Westbury. He was the tenth who died. His aged father, crying that there was a Curse, threw himself from an upper window, and so become the eleventh. At his funeral, the hearse ran over a boy of eight years, and so killed the twelfth. At the British Museum, Edgar Steel, aged fifty-seven years, after handling certain of the relics from the tomb, was mentioned as the thirteenth. . . .

"Historians tell us that the aristocrats of ancient Egypt were experts on immortality. They were most careful, therefore, to be buried, at once with the utmost splendor and the utmost secrecy. With such safeguards, indeed, the Egyptians did not rest satisfied. There was added what, in the language of the occult, is called a Curse . . . levelled at anyone who at any time should be so impious as to disturb the sleeper and rob him of his possessions. Over every such tomb, so it was firmly believed, there brooded an unseen guardian, charged with the duty of enforcing the solemn imprecation."

The striking fact is the thirteen deaths. What was behind it? You have just read one possible explanation:

H: An Egyptian god, fulfilling the contract of the Curse, caused the deaths.

Others believed:

H_c: Blind chance caused the deaths.

It is easy to make a convincing argument that H_c is true. Thirteen deaths is a modest occurrence, considering that many hundreds of people were involved in excavating the tomb and archiving the finds. Base-rate life table data make it no surprise that a number of people would die. What's more, any vengeful Egyptian god surely would have struck down Howard Carter, yet he lived into his sixties. Or look at it this way: There are about 1,000 active and retired faculty members at my university. Over any two or three years it is not unusual that at least 13 of them die. Is my university cursed?

Res. Ex. 6.18: Does the Bible contain hidden prognostications?

With computer spadework, Drosnin (1997) dug some remarkable facts out of the Hebrew Bible, which according to legend was given to Moses by God. Omitting the punctuation and removing the spaces between words, Drosnin put the Bible's text into a computer and arranged the resulting string of letters into a matrix. He programmed the computer to search the matrix vertically, horizontally, and diagonally for series of letters that spelled keywords that he specified. Next, he searched in the text for prophetical words located near the keywords, and almost invariably he found them. To illustrate, one keyword he specified was *Kennedy* (for President Kennedy). The computer found a string of letters that spelled *Kennedy*, and near it was the prophetic word *Dallas*. In the end, Drosnin uncovered many instances of various keywords linked prophetically to nearby words. Wishing to explain this miraculous fact, he retroductively hypothesized:

H: God embedded a prophetic code in the Bible.

Drosnin concluded that H is true because he saw no other possible explanation.

It remained for Thomas (1997) to propose a possible explanation:

H_c: What is called the Bible's prophetic code is the effect of chance.

To test H_c, Thomas needed base-rate data of the frequencies at which words occur in large matrices of texts. To generate it, he put two texts into a computer and arranged each in a matrix: the text of the King James Version of the Book of Genesis, and of the U.S. Supreme Court's 1987 ruling on creationism. Next, he specified keywords, and the computer found them in the texts. And almost always near the keywords were sequences of letters that spelled words related to the keywords. For instance: In both texts, near the keyword *Nazi* was the word *Hitler*, and near *UFO* was *Roswell*. Moreover, the word *Hitler* appeared 18 times in the Bible, and *Stalin* appeared 15 times. In the Supreme Court's ruling, *Hitler* appeared 13 times and *Stalin* 10 times. Both texts contained instances of the words *UFO*, *alien*, *saucer*, and *Roswell*.

The upshot is that a good many meaningful words are expected to occur in matrices made of any long text, such as an issue of *The New York Times*, an automobile shop manual, or even several hundred thousand random letters of the alphabet. No doubt about it, the Bible's hypothesized prophetic code is no more miraculous than chance is miraculous.

Res. Ex. 6.19: What about the geometry of the sun and the moon?

The sun's diameter is 400 times the moon's, and the sun is 400 times as far from Earth as the moon. This makes the sun and the moon appear the same size to us. In turn, this gives us striking total eclipses, where the sun's corona radiates sublimely. Science provides no physical reason for why the special geometry exists. But there's nothing to stop believers in miracles from providing a supernatural cause:

H: God arranged the sun-moon geometry to offer us striking eclipses of the sun.

So what are scientists to do? Test the hypothesis of chance:

H_c: The sun-moon geometry is just a coincidence.

For testing H_c, we have to be careful that the base-rate data we use include all the kinds of solar-system facts that we would consider striking. Otherwise, the test will be wrong. A simple illustration of a wrong test will make this clear: In total, the planets have more than 100 moons. Excluding Earth, if we could stand on each of the other planets that have moons and look at the sky, the apparent sizes of the sun and each of the planet's moons would not be the same (so far as we know). If we take this for base-rate data, Earth's striking sun-moon geometry exists in 1 percent of the cases. By this we would think that H_c is most likely false, and by that conclude that God caused the sun-moon geometry. But wait; the base-rate data are overly restrictive. They contain only one kind of striking solar-system fact: sun-moon geometries.

To correctly test the hypothesis of chance, the base-rate data must allow for all kinds of striking solar-system facts. There are plenty. Chance easily delivers at least one "wowing" event per planet. Among them: Oddly enough, Jupiter has its Great Red Spot (Wow!). Saturn has Iapetus, a walnut-shaped moon, half white and half black, with a 20-kilometer-high ridge around its middle (Wow!). Mars is red (Wow!). Venus spins backwards, making the sun rise in the west and set in the east (Wow!). Mercury is hottest (Wow!). Judging from this base-rate, it's a good bet that our sun-moon geometry is simply a randomly determined condition.

Step 3 for trying to debunk a claim of a supernatural cause. We reach step 3 whenever we are unsuccessful at step 2, either because the test of the hypothesis of chance failed to eliminate chance as the cause, or because no suitable base-rate data existed to do the test. At step 3, then, we apply Littlewood's method of testing the hypothesis of chance. The method rests on Littlewood's Law of Miracles, which specifies a general base-rate for miracles. As Dyson (2004) explains:

> "[Miracles] are not the result of deliberate fraud or trickery, but only of the laws of probability. The paradoxical feature of the laws of probability is that they make unlikely events happen unexpectedly often. A simple way to state the paradox is Littlewood's Law of Miracles. Littlewood was a famous mathematician who was teaching at Cambridge University when I was a student. Being a professional mathematician, he defined miracles precisely before stating his law about them. He defined a miracle as an event that has special significance when it occurs, but occurs with a probability of one in a million. This definition agrees with our common-sense understanding of the word 'miracle.'
>
> "Littlewood's Law of Miracles states that in the course of any normal person's life, miracles happen at the rate of roughly one per month. The proof of the law is simple. During the time that we are awake and actively engaged in living our lives, roughly for eight hours each day, we see and hear things happening at a rate of about one per second. So the total number of events that happen to us is about thirty thousand per day, or about a million per month. With few exceptions, these events are not miracles because they are insignificant. The chance of a miracle is about one per million events. Therefore we should expect about one miracle to happen, on the average, every month."

Let's look at a case of applying Littlewood's Law of Miracles to debunk the claim of a supernatural cause. Suppose someone in Houston cuts a branch from a tree and the sap at the cut site oozes into what looks like the weeping face of Mary, mother of Jesus Christ. Word travels; people soon make it a shrine, saying it is a miracle caused by God. How to rationally counter this?

Step 1 fails: natural laws cannot explain sap drawing the face of Mary. Step 2 fails: no specific base-rate data exists for a test. Therefore, we apply Littlewood's method of testing the hypothesis of chance, which compares the miracle of the weeping face to the general base-rate of miracles. By Littlewood's Law, the weeping face is just one of those ordinary miracles that appear every now and then, which argues against its having a supernatural origin.

7

How to discover knowledge by building and using reasoning models

I know something about you. Very probably, some of the water you drank this morning was drunk by Aristotle. I didn't discover this with any of the research methods presented so far in this book. I did it with a reasoning model. I started with several facts, and from them I predicted, by deductive reasoning, the fact that you drank some of Aristotle's water this morning.

I began by considering an arbitrary 20-year period of Aristotle's life, 7,300 days. I estimated the fact that each day he drank at least 32 ounces of water, or 233,600 ounces in that period. I looked up another fact that there are 43×10^{20} molecules in one ounce of water. Hence, Aristotle drank at least $233,600 \times 43 \times 10^{20} = 10^{27}$ molecules of water within the 20 years. Most of these molecules are now mixed in the water of the world's oceans, lakes, reservoirs, and rivers. I looked up the fact of how much water this is and found out that it would fill about 4.7×10^{22} 8-ounce glasses. I divided the 10^{27} molecules that went through Aristotle by 4.7×10^{22} 8-ounce glassfuls. This told me that each glassful contains on average 21,300 molecules of Aristotle's water. Whenever you drink 8 ounces of water, you can hardly escape getting some of his water (not to mention Cleopatra's, Confucius's, and Christ's). Dawkins (1998) attributes the logic of this deduction to Lewis Wolpert; the calculations illustrating it are mine.

Five things will strike you about the research examples of reasoning models in this chapter. One is that the input to the reasoning models consists of a few facts. The facts are qualitative or quantitative. If qualitative, scientists believe them without question. If quantitative, they are reliably estimated, acceptably free of random error and systematic error. The second thing you will notice is that the output of the models is a predicted fact, which may be qualitative or quantitative. The third thing you will notice is that the reasoning from the input facts to get the predicted fact is deductive. One can do no better than that: reliable facts operated on with reliable reasoning results in a reliable predicted fact. The fourth thing you will notice is that the equipment required to build a reasoning model and make a prediction from it is no more than pencil, paper, and in some cases a pocket calculator. The last thing you will notice is that most reasoning models are the work of one person, not a team.

Fig. 7.1 is a P-plane diagram for the reasoning model for "Aristotle's water." Its input consists of three reliably estimated quantitative facts. F_1 is the amount of water Aristotle drank. F_2 is the number of water molecules per glassful. F_3 is the number of glassfuls of water in the world. From these facts, the deduction goes across the C-field above the P-plane, predicting the quantitative fact, F_p, the average number of Aristotle's water molecules that today are in each glassful of water. We cannot measure the experimental fact, F_e, from the P-plane. In its place, we accept the predicted fact, F_p. In doing so, we don't settle for second best. F_p is reliable. I would wager thousands of dollars (wouldn't you?) that the prediction about your drinking Aristotle's water is correct.

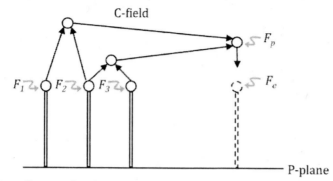

Figure 7.1. A P-plane diagram for the reasoning model for "Aristotle's water."

Food for your thoughts: The term "reasoning model" makes it convenient to present deductive reasoning according to its purpose. (1) The purpose of one kind of reasoning model is to predict a fact we want to know from one or more facts we do know. (2) The purpose of another kind of reasoning model is to predict a fact not because we want to know what the fact is (sometimes we already know it), but because we want to explain nature's process that connects the one or more facts we know to the predicted fact. (3) The purpose of the final kind of reasoning model is to do thought experiments: we assume a fact or facts, and from them we predict a fact to discover the consequence of our assumption. The coming sections 7.1, 7.2, and 7.3 cover the three.

Further than this, the term "reasoning model" is parallel to the term "computer simulation model." Computer simulation models, the subject of chapter 8, handle the same three purposes.

7.1 Using reasoning models to predict facts

Here we take one or more known facts and from them predict a fact we want to know. It may be that the fact we want to know is too costly to obtain by experiment. Or it may be impossible to obtain by experiment, as with the case of Aristotle's water.

Let me show you a variety of reasoning models for predicting facts, beginning with a reasoning model a child could make.

Res. Ex. 7.1: How many hekats of grain is saved?

Reasoning models go back to ancient times. Newman (1956) cites one written on papyrus, interpreted as: "In each of 7 houses there are 7 cats; each cat kills 7 mice; each mouse would have eaten 7 ears of spelt; each ear of spelt would have produced 7 hekats of grain. How many hekats of grain is saved by the cats?" From these input facts, the predicted fact is 16,807 hekats (7 x 7 x 7 x 7 x 7).

As we continue with more research examples, words like "deduction," "calculation," and "predicting" signify that a reasoning model is being used.

Res. Ex. 7.2: How many cubic meters of stone are in the Great Pyramid at Giza?

Bernard Cohen (2005) describes how Napoleon deduced the amount: "While the more adventurous officers climbed to the top, Napoleon himself was content to rest in the shade of the pyramid at its base, toying with numbers. When the officers descended and rejoined him, Napoleon announced that he had made a calculation of the amount of stone in the pyramid. There was enough, he said, to build a stone wall 3 meters high and 0.3 meters thick (about 10 feet high and 1 foot thick) that would enclose the whole of France. One in the French group was Gaspard Monge, a great mathematician and the inventor of projective geometry. He made his own calculation and is said to have declared that Napoleon had been quite right in his estimate."

For the moment let's stay in Egypt:

Res. Ex. 7.3: How many workers built the Great Pyramid at Giza?

Stuart Kirkland Wier deduced the answer (Stewart 1998). First, from known facts he calculated the energy required to quarry the pyramid's blocks and transport and lift them to their heights. Second, he divided the energy required by the estimated time it took to build the pyramid, 23 years, or 8,400 days. This gave the average energy required per day. He divided this by the fact of the amount of energy an average worker was capable of supplying per day. He found that about 10,000 workers would have been enough. Think of it! The reasoning model does away with the need to hop into a time machine and go back and count workers.

Let's leave ancient Egypt for the far more ancient Cretaceous era.

Res. Ex. 7.4: Where did the pteranodon live?

Though we can't peek into the remote past to find out, Tennekes (1996) has deduced that the anatomy of pteranodons restricted them to living along ocean cliffs. The facts we have to go on are implied by fossil remains collected along the shoreline of California, dated to the Cretaceous era. A typical pteranodon weighed 37 pounds, had a wingspan of 23 feet, and a wing area of 108 square feet. To fly with this ratio of weight to wing area, Tennekes deduced that it would have had to flap its wings continuously at a faster rate than its size indicates it was capable of. Thus it needed wind assistance. Yet there was little global wind at the time. How do we know? From another deduction. From the fact that in the Cretaceous era there were no polar ice caps, Tennekes deduced that the temperature difference between Earth's poles and its equator was quite less than it is today. Because this temperature difference drives global winds, the wind that pteranodons needed to stay aloft had to be local. And local wind was dependably available only along ocean cliffs.

Tennekes finished his deductions with this prediction about the past: "A pteranodon spent its life soaring above the cliffs along the shoreline, since it was not nearly strong enough for continuous flapping flight. Its airspeed was about 7 meters per second (16 miles per hour) – not fast for an airborne animal that must return to its roost in a maritime climate."

Res. Ex. 7.5: Do reservoirs affect the length of a day?

Scoop up trillions of tons of water from the Northern Hemisphere's low latitudes and put it in the middle latitudes where it is closer to Earth's axis of rotation. As happens to spinning ice skaters who pull in their arms, Earth will rotate faster. Such a northward shift of water has occurred in the past 40 years. To supply water to the Northern Hemisphere's population, reservoirs containing about ten trillion tons of water were built.

From the facts of where the reservoirs are located and of their sizes, Chao (1995) deduced that the length of a present day is 8 millionths of a second shorter than it was 40 years ago. And from the fact that the reservoirs are lopsidedly situated about Earth's axis, he deduced that this has shifted Earth's axis of rotation about 60 centimeters away from where it was pointed 60 years ago. Making reservoirs, he concludes, is "the only human activity that is able to cause appreciable changes to Earth's rate of spin, and to the direction its axis points."

Res. Ex. 7.6: How much will the completed Three Gorges Dam warm Japan?

Scheduled for completion before 2010, the dam on China's Yangtze River will create a reservoir as long as Lake Superior. According to Perkins (2001), a researcher named Doron Nof deduced some of the probable consequences if as little as 10 percent of the dammed water goes for irrigation. That will reduce the water flowing into the sea. In turn, the sea's surface will become saltier and sink. In turn, warm water will come to the surface. In turn, the air will be heated. In turn, the air temperature in the region of Japan will rise several degrees Celsius.

Res. Ex. 7.7: What happens to the oceans in a greenhouse-warmed world?

Johannes Weertman deduced a chain of consequences. The west Antarctic ice sheet will melt. All the oceans will rise by 5 meters. New Orleans and Bangkok are just some of the cities that will be under water. The bottom third of Florida will be submerged. The rate of flooding won't be constant; it will increase, Kerr (1998) reports: "[Weertman] explained that even a slight warming-induced retreat of the ice's grounding line – where it begins to float off the bottom – will move the grounding line into thicker ice. The thicker the ice, the faster it flows outward and therefore the faster it thins. The faster it thins, the sooner it floats and moves the grounding line even farther inward. Such an accelerating retreat could consume the West Antarctic ice sheet in a matter of a century or two. . . ."

The next two research examples illustrate that the facts predicted with reasoning models can be probabilities.

Res. Ex. 7.8: How likely is it that the Sun and another star could collide?

Here are three facts: (1) The Sun is estimated to be one of a hundred billion stars in our galaxy, all milling around. (2) The Sun has enough remaining fuel to last five billion years. In view of this, we would not be laughed out of town for thinking that someday the Sun is likely to collide with a star. Yet another fact belongs in the deduction: (3) Our galaxy is trillions and trillions of times more spacious than the space occupied by its stars. There is enormous room for stars to miss bumping into one another. When this is considered, the calculated probability of the Sun colliding with another star is vanishingly small. Tyson (1989) gives an idea of the improbability this way: "If there were just 48 bumble bees randomly buzzing throughout the United States then it is more likely that two of them will accidently bump into each other than it is for another star to collide with the Sun."

Res. Ex. 7.9: How probable is life in red dwarf solar systems?

Of all types of stars, red dwarfs are commonest. They are cooler and older than the Sun. Because they are cooler, the current habitable zone around them, where temperatures are suitable for life, is closer to their surface than the habitable zone around the Sun is to its surface. But is there life on planets in red dwarf habitable zones? Probably not. Dr. Jack J. Lissauer, National Aeronautics and Space Administration, Ames Research Center, has deduced that there is probably no water or other volatiles on planets there. Deduction 1: Billions of years ago, red dwarfs were far hotter than now. Planets in the habitable zone would have been scorched, boiling away their water and volatiles. Deduction 2: Because the solar system disks around red dwarfs do not stretch out far, they are dense with asteroids and rock debris. Planets would have been struck many times, this too causing water and volatiles to be lost.

Note the one-two punch. The two deductions constitute two reasoning models that predict the same fact: water is improbable on planets circling red dwarfs, and thus life is improbable.

The painter Emile Nolde (1867-1956) once argued that it is highly likely that intelligent life is abundant in our galaxy. Quoting from Chipp (1968), Nolde asked: "Who is conceited enough to assume that our earth is distinguished among the planets? What is so significant about the earth, this little thing, in the universe of millions of far greater bodies where, measured by earthly standards, the chances for life and development are incomparatively better?" But why stop there? The sequel to Nolde's argument is that intelligent life in our galaxy is attempting interplanetary communication.

Res. Ex. 7.10: How many communicative civilizations exist in our galaxy?

We can get a prediction from an equation named for cosmologist Frank Drake. The Drake Equation (Nadis 2002) is the product of seven factors: (1) the number of stars like the Sun created each year in the Milky Way; (2) the fraction of these stars that have a solar system of planets; (3) the number of habitable planets per solar system; (4) the fraction of habitable planets that develop life of any sort; (5) the fraction of these planets where the life is intelligent life; (6) the fraction of planets with intelligent life that use technology to communicate; (7) the average number of years a communicative civilization lasts. Making conservative estimates of the seven factors and multiplying them together, it comes out that there are several million communicative civilizations in the Milky Way.

Very well, but what if a fact needed for input to a reasoning model doesn't exist? Possibly it can be measured, becoming known, as the following shows:

Res. Ex. 7.11: How did Ben Franklin predict the size of a molecule of oil?

As reported in Volume 64, Part 1 (1774) of *Philosophical Transactions,* the journal of the Royal Society, Ben Franklin measured out a volume of 5 thousandth of a liter of oil and poured it into a pond. It spread into a slick, which he reasonably assumed was one molecule thick. He estimated the slick's area to be 2,000 square meters. With that, he calculated what the thickness of the slick would be to make its volume be 5 thousandth of a liter. He got 2.5 billionth of a meter. That is almost exactly right. Measurements today with electronic instruments put the diameter of a molecule of oil at about 2 billionth of a meter.

While we're pouring liquids:

Res. Ex. 7.12: How to pour a thin layer of mercury?

The larger the mirror of a reflecting telescope, the more light it gathers. But large glass mirrors are expensive to cast and polish. The 6-meter telescope at the Liquid Mirror Observatory in Canada is cheaper. Its mirror is a 6-meter diameter dish of mercury shaped in a 1-millimeter-thick parabolic layer by continually spinning the dish at about 7 times per minute. The telescope can only look straight up, but from that limited angle there is much to see in deep space.

The mirror's 30 liters of mercury get dusty and must be drained and cleaned every few weeks. The problem in refilling it is that the mercury's surface tension prevents pouring a 1-millimeter-thick layer. Engineers reasoned out a way to defeat the surface tension. Pour 60 liters of mercury into the dish to overcome the balling-up effect of surface tension, and drain off 30 liters, leaving a 1-millimeter-thick layer. This worked as predicted (Schilling 2003).

Res. Ex. 7.13: Predicting your car's weight

You'll need to collect some facts to go into the deduction. For that, at your disposal is a ruler, a bucket of ice, a tire gauge, a flame thrower, a pencil, ten pieces of paper, and a box of gumdrops.

First, put a piece of paper on smooth pavement. Push the car so that its left front tire is on the paper. With the pencil, trace out the tire's rectangular "footprint" and compute its area in in^2. With the tire gauge, measure the lbs per in^2 pressure in the tire. Repeat this for the other three tires. After that, multiply each tire's footprint area by its pressure, to get the number of pounds each tire is supporting. Add these pounds for the four tires, and that is an estimate of your car's weight. Swartz (2003) did this for a 1991 Lexus. He got 3,120 lbs, close to 3,133 lbs, the weight according to the manufacturer.

Comment: Problem solving in school is much easier than problem solving in real life. In school, students are generally supplied exactly what is needed to solve a problem. In real life, they have to select what is needed from a profusion of choices. When I have posed to students the problem of deducing a car's weight, complete with the red herrings – the bucket of ice, the flame thrower, and the box of gumdrops – few have solved it. When I have posed it to other students, offering just the required equipment of pencil, paper, ruler, and tire gauge, more of them got it.

Comment: Provided you have a Saturn rocket, you can deduce your car's weight with another kind of reasoning model. Take your car into outer space. Apply a defined force, *F*, to it, and measure its resulting acceleration, *a*. By Newton's second law of motion, $F = ma$, solve for your car's mass, *m*. Then convert its mass to its weight on Earth.

7.2 Using reasoning models to explain a fact

Here we build a reasoning model that deduces a fact. The steps of the deduction are, it follows, the explanation that accounts for the deduced fact.

Prized explanations have come this way. I'll tell you one of them. The research of Francis Crick and James Watson. They created a reasoning model to explain the structure of DNA. The input facts were (1) the molecular structures of four nucleotides, (2) a rule of chemistry called Chargaff's Law about the relative concentrations of the four nucleotides, and (3) the fact of the helical structure indicated on an X-ray diffraction photo of DNA. From these inputs, they deduced how paired nucleotides and their companion molecules can be stacked to make a double helix. With that, it burst upon them that "unzipping" the double helix explains how DNA could replicate.

Let's now move on to less famous applications of reasoning models to explain why things are as they are.

Res. Ex. 7.14: Why do adults go Ouch! when they walk barefoot on pebbles?

One fact is that you weigh many pounds more today than when you were five years old. A second fact is that when you walk on pebbles now it hurts you much more than when you were a child. From the first fact Tennekes (1996) has deduced the second, and by that explained the physical process connecting the first to the second. It's that as you grew up, your volume and thus your weight increased disproportionally to the increase of the area on the bottom of your feet. When you step on pebbles barefooted now, there are more pounds per square inch at the pressure points of the pebbles than when you were little. For barefoot adults to walk as painlessly on pebbles as they did as children, they would have to have grown feet like Donald Duck's, with more bottom surface area.

In the same way, we can deductively explain why a sharp knife cuts more easily than a dull knife. Sharpening decreases the blade's cutting area. Apply a force to a sharpened knife, and the force is concentrated on a smaller area. Likewise, we can deductively explain why it hurts less to be stepped on by someone wearing jogging shoes than someone wearing stiletto heels.

I want to pass on to you an educational tip about a good way of learning to build and use reasoning models. Instead of learning each reasoning model as a separate case, based on its separate flesh, learn the far fewer kinds of common bones. I want to drill this home with a dozen or so examples, all with different flesh on common bones. The common bones in this case involve spheres.

Of all three-dimensional shapes, spheres have the smallest ratio of surface area to volume. This can be proved mathematically. Or you can demonstrate it to yourself experimentally by taking about eight cubic inches of sugar and spreading it out thinly on a piece of plastic wrap. The sugar has a top area, a bottom area of the same size, and a bit of area on its side. Added up, these areas are the sugar's surface area. Next, gather the four corners of the plastic wrap, pass them through a napkin ring, and pull the corners up. As you pull, you are reducing the sugar's surface area while its volume remains the same, eight cubic inches. When you can pull no more, you have reduced the surface area to its limit. What's the shape? A sphere.

Besides, when you were a child you demonstrated the same thing when you blew soap bubbles. You never blew a bubble that was a cube or a pyramid or an octahedron. The bubble's surface tension gave it the smallest possible surface area for its volume, a sphere.

Res. Ex. 7.15: Why do you curl up in bed when you are cold?

It is through your surface area that you lose heat to the cooler surrounding air. As you curl up, your volume remains constant and your surface area reduces, so you lose heat at a lesser rate. In curling up you are trying to mimic a sphere, to the extent your anatomy allows. Domestic cats, being more flexibly jointed than humans, curl up more in a ball, closer to the shape of a sphere, the better to conserve body heat.

The same reasoning model is the bones for reasoning about teapots and snakes:

Teapots are approximately globular. When it comes to heat retention, the optimal shape for a teapot is a sphere. Long skinny teapots quickly lose heat. The worst imaginable teapot would be adapted from a car's radiator.

Snakes act to maintain their body temperature within a narrow range. Their long string of joints allows them infinite degrees to adjust their surface area, regulating their exposure to sunlight. In coiling and uncoiling, their volume remains constant. What changes is their surface-to-volume ratio. A snake coiled has a smaller surface-to-volume ratio than the same snake uncoiled.

Comment: You've just seen the difference in answering a question in terms of knowledge of cause-and-effect, and in answering it in terms of knowledge of processes. When I asked, "Why do you curl up in bed when you are cold?," you could have answered with "to keep warm." That answer is correct in terms of cause and effect. The cause, curling up, produces the effect, keeping warm. Instead, we answered the question with a reasoning model in terms of the process of how heat transfer is changed as surface area is changed.

Res. Ex. 7.16: Dog droppings and potato chips

The fact is that a dog dropping smells more strongly after it is stepped on. Deduction tells us why. Round or squashed out, its volume is the same. Stepping on it increases its surface area, increasing the density of smelly molecules in the air. The odor is strongest when the ratio of fecal surface area to the given fecal volume is largest. In a word, diarrhea.

As to potato chips, it's on purpose that they are potato chips, not potato spheres. Potato chips are made to have large surface areas relative to their volumes. To get the most taste buds in contact with the chip's yummy fat, it's best to have the potato material squashed out.

Before going on, I want to go through a derivation of a law about spheres. With D as the diameter of a sphere, make a ratio, R, of the sphere's surface area, $A = \pi D^2$, to its volume, $V = (\pi/6)D^3$. After cancelling what can be cancelled, the resulting ratio of surface area to volume is $R = 6/D$. This is the law, and it tells us that the larger D is, the smaller R is, an inverse relation. And although it cannot be shown with an equation, it's true for non-spherical objects, such as cylinders, cones, and cubes, and for geometrically irregular shapes like a lump of clay, that the larger the object is, the smaller is the ratio of its surface area to its volume. We are ready to make use of this.

Res. Ex. 7.17: Why are the mammals in cold climes big and rotund?

Whoever saw a tiny twiggy polar bear or walrus? Low temperature in the Arctic's winter goes with big rotund mammals. Rotundness approximates sphericalness. It's nature's way of packing lots of flesh into a given surface area, which minimizes heat loss. And the bigger the sphere, the further it minimizes heat loss.

Please read this extract from J. B. S. Haldane's essay, "On Being The Right Size" (Haldane 1928): "One of the most obvious [advantages of size] is that it enables one to keep warm. All warm-blooded animals at rest lose the same amount of heat from a unit area of skin, for which purpose they need a food-supply proportional to their surface and not to their weight. Five thousand mice weigh as much as a man. Their surface and food, or oxygen consumption, are about seventeen times a man's. In fact a mouse eats about one-quarter its own weight of food every day, which is mainly used in keeping it warm. For the same reason small animals cannot live in cold countries. In the arctic regions there are no reptiles or amphibians, and no small mammals. The smallest mammal in Spitzbergen is the fox. The small birds fly away in the winter, while the insects die, though their eggs can survive six months or more of frost. The most successful mammals are bears, seals, and walruses."

Comment: Haldane's deductively reasoned explanation of why arctic mammals are large isn't a hypothesis. It is correct beyond dispute. That is because deduction, provided we don't goof and make an error of logic in doing it, has to be correct.

Res. Ex. 7.18: Why bundle up children in cold weather and feed them plenty?

The answer is in the above quotation (Res. Ex. 7.17) of J. B. S. Haldane. To wit, children's bodies, relative to adults', have greater surface area per body volume, and thus per pound of weight. Consequently they need more food per pound of their weight than adults need per pound of theirs. Huxley (1929) speaks to the point: "The reason that children need proportionately more food than grown-ups is not only due to the fact that they are growing, but also to the fact that their heat loss is relatively greater. A baby of a year old loses more than twice as much heat for each pound of its weight than does a twelve-stone man [168 pounds]. For this reason, it is doubtful whether the attempt should be made to harden children by letting them go about with bare legs in winter; their heat requirements are greater than their parents', not less."

This holds equally for domestic animals outside in the cold. The smaller the animal, the more urgent it is to provide a place of shelter, and to feed it relatively more food. Suppose you have two outdoor dogs, the smaller weighing half as much as the larger. It follows by deduction that you should feed the smaller one something more than half of what you feed the larger one. The smaller one, with its larger surface-to-volume ratio, loses heat more quickly.

Res. Ex. 7.19: Why do Siberians' heads tend to the spherical?

Fact 1: Siberia gets cold. Fact 2: The temperature of our brains must be kept within a narrow range. From facts 1 and 2 we can deduce fact 3: The heads of Siberians ought to be broader, more to the spherical, than the heads of, say, Italians. For, as the steps of the deduction go, the more a head conforms to a sphere, the smaller is its surface-to-volume ratio, and so the better it is for insulating it against heat loss. As it happens, the actual fact 3 is available to test this explanation against (not that we need to). Listen to Balter (2005): "Anthropologist Charles Roseman of Stanford University in California last year reported [in the *Proceedings of the National Academy of Sciences*] that the skulls of the Buriat people of Siberia are broader than predicted by random drift. Broad skulls have smaller surface areas and so may be an adaptation to cold climates." The same has been found for Greenland Eskimos.

Res. Ex. 7.20: Why do hummingbirds have relatively small red blood cells?

Fact 1: Hummingbirds are small. Fact 2: Hummingbirds have small red blood cells. From the first fact, scientists have deduced the second fact, and the steps of the deduction amount to the explanation of why hummingbirds have small red blood cells. As we go through the steps, be aware that the surface-to-volume ratio enters twice:

Among bird species, the hummingbird is small, giving it a large surface-to-volume ratio. Thus it loses heat at a relatively high rate. Thus it must almost continuously eat to supply its metabolic furnace. At the same time, its metabolic furnace needs oxygen fast to burn its food's calories fast. Thus the hummingbird's red blood cells need to capture oxygen molecules fast in its lungs, and it needs to transport the captured oxygen fast across its cell membranes. Thus its hemoglobin molecules are packaged in small, not large, red blood cells. Small red blood cells have relatively more surface area for oxygen molecules to attach to as the red blood cells pass through the lung. And small red blood cells more readily slip through cell membranes, headed for the metabolic furnace. This explanation generalizes. Species of small mammals and small birds tend to have small red blood cells.

Res. Ex. 7.21: Grain elevator explosions and crushed ice

Put a hockey stick on a fire and it burns slowly. Mill a hockey stick to dust the size of talcum, blow the dust into the fire, and hold on to your hat. In both situations, the stick gets burned up, releasing equal amounts of energy. But in the second, the energy is released rapidly.

Milling a volume of wood or grain into small particles leaves the volume unchanged while creating surface area. When a small particle ignites, it burns so fast it ignites several nearby particles. In turn, they burn so fast they ignite nearby particles. It's a chain reaction at an ever-increasing rate, a runaway.

On June 8, 1998, in Wichita, Kansas, a grain elevator exploded, killing seven. According to the U.S. Occupational Safety and Health Administration's Executive Summary report, "witnesses reported that during elevator operation that the cloud of suspended grain dust was often so thick that during these times one could not see their hand in front of their face. . . . The most probable ignition source was created when a concentrator roller bearing, which had seized due to no lubrication, caused the roller to lock into a static position as the conveyor belt continued to roll over it. This 'razor strop' effect on the roller raised its temperature to 260° C, well beyond the 220° C required to ignite layered grain dust which was plentiful inside the roller."

Keep the bones of this reasoning model the same but change its flesh from dust to atomized liquid gasoline, and we have the reason behind fuel injection systems. Atomized gasoline burns faster than larger droplets. Or change the reasoning model's flesh to ice. Make tea, let it reach room temperature, add three ice cubes, and start your stopwatch, timing how long it takes the tea to cool down. Repeat the experiment, but this time crush the three ice cubes, increasing their surface area, and we would predict that the tea will cool faster. (Question: Why do bath towels have plenty of nap?)

So far the research examples involving spheres have considered individual entities in isolation: a potato chip, a child, a particle of grain, a polar bear, a snake. . . . It is time to look at collections of individuals that form groups.

Res. Ex. 7.22: Why snuggle and huddle in the cold?

On warm nights, bed-mates will sleep apart. On freezing nights, the two will make themselves cozier by nestling in to each other. With the front side of the one pressed to the backside of the other, the surface area of the two is reduced from what it would be with the two apart, while the total volume of the two is unchanged. Snuggling, the two have a smaller ratio of surface area to volume, minimizing their heat loss. (On hot summer nights, when the two want to shed heat, they increase their surface area by moving apart.)

Under other circumstances, the fact is that during the long Antarctic winter, male Emperor Penguins stand close together in a circle. Taking turns, each spends a few minutes standing at the perimeter before moving inside for warmth. Why a circle? Of all two-dimensional shapes, a circle has the smallest ratio of perimeter to the area it encloses. A circle is the shape that minimizes the loss of heat through the group's perimeter.

Another fact is that the penguins group themselves in one big circle rather than in a number of smaller circles. The fact is deductively explainable. With the penguins in one circle, the ratio of the perimeter to the enclosed area is some number, R. With the penguins in two circles of the same or different sizes, the ratios of the perimeters to the enclosed areas are R_1 and R_2. It can be shown algebraically that R is less than $R_1 + R_2$. This makes the heat loss of the large group less than the total heat loss of the two smaller groups.

We are nearly at the end of this section. I may have left you wondering, "Isn't a reasoning model for explaining a fact just an instance of retroduction, as covered in chapter 3?" In one sense it is. Reasoning models explain facts, and retroduction explains facts. Beyond that lies a main difference: With retroduction, the research hypothesis we create is a plausible explanation of a fact. It has the possibility of being right or wrong. Whereas with a reasoning model to explain a fact, the explanation we create is not a hypothesis. It is correct. There is no doubt about it because it is deduced from one or more correct starting facts, using if-then propositional logic, or if-then mathematical logic, or if-then logic based on physical law(s). Look here, the explanation of why you curl up in bed on cold nights (Res. Ex. 7.15) is correct, and could only be incorrect had we made a mistake of logic, which we didn't.

I hope I haven't left you with the impression that surface-to-volume ratios play a majority role in reasoning models. They don't. I went overboard on them to emphasize that one type of deductive bones can be fleshed out in different ways for different situations. You will get a more balanced view by reading books featuring reasoning models. Two good ones in biology are Stephen Vogel's *Life's Devices* and *Cats' Paws and Catapults*. They are packed with reasoning models that have nothing to do with surface-to-volume ratios. The reasoning models draw on laws and principles of all sorts of subjects: aerodynamics, hydrodynamics, strength of materials, optics, surface tension, osmosis, pumps, fibers, metals, and more. And a good book for its variety of samples of reasoning models in physics is Clifford Swartz's *Back-of-the-Envelope Physics*.

7.3 Using reasoning models to do thought experiments

Thought experiments are experiments in an imaginary world. They amount to posing a hypothetical, like "What if this certain imagined view of nature is true?", or "Assume that such is the case." To run thought experiments, we deductively reason out the consequences of what we imagine. Sometimes the consequences reveal knowledge of how nature is. Other times they reveal knowledge of how nature is not. Either way, we learn something of nature.

Freeman Dyson (1979) has put this in different words, writing that a thought experiment is "an imaginary experiment which is used to illuminate a theoretical idea. It is a device invented by physicists; the purpose is to concoct an imaginary situation in which the logical contradictions or absurdities inherent in some proposed theory are revealed as clearly as possible. . . . [It is] a tool for weeding out bad theories and for reaching a profound understanding of good ones."

In German, thought experiments go by the name of "gedanken experiments" (gedanken is German for thought). Einstein famously performed gedanken experiments to discover physical truths that led him to create the Special Theory of Relativity.

Why should what we discover in an imaginary world be valid in the real world? To begin with, we put in the imaginary world the same laws of nature that exist in the real world. The reasoning we use with these laws of nature in the imaginary world follows the same rules of logic as the reasoning we use with these laws in the real world. The imaginary world's difference with the real world is only with things that don't matter to our reasoning in the imaginary world. No matter then that we have to put, say, a 100-mile high mountain in the imaginary world, as part of the imagined equipment for a thought experiment. This is all right as long as the deductive conclusions we reach with the thought experiment do not depend on the presence of a 100-mile high mountain in the real world.

Let me drill this home with a selection of research examples of thought experiments, beginning with one of Galileo's:

Res. Ex. 7.23: In a perfect vacuum, do heavy objects fall faster than light objects?

Like Galileo, we lack equipment for producing a perfect vacuum and for errorlessly measuring and recording the time it takes objects to fall a fixed distance. But like him, we can do a thought experiment to discover the answer.

Imagine 50 marbles weighing exactly the same. Imagine they are dropped simultaneously from a height of 100 feet at exactly the same time in a perfect vacuum. Deduction: All will reach the bottom at the same time.

Imagine the same experiment except the 50 marbles are connected by a piece of weightless string so short that the marbles are gently touching, making a mass 50 times as heavy as a marble. Deduction: The mass of marbles will reach the bottom at the same time as the marbles in the first imagined experiment did.

Together, the thought experiments imply that an object's weight has nothing to do with its rate of acceleration in free fall. Notice that: (1) Reasoning out the answer takes a tiny fraction of the cost of doing the real experiments. (2) Reasoning out the answer is errorless; the measurements for the real experiments would not be. (3) Reasoning out the answer gives a universal conclusion. Release Rolls Royces or single atoms of chromium from a tower on the asteroid Ceres, and the released objects will race downward exactly neck and neck.

Res. Ex. 7.24: Why doesn't the moon fall out of the sky onto Earth?

As one version of the story goes, Newton imagined a mountain hundreds of miles high. He imagined a really strong person on the peak. He imagined the person throws a stone horizontally and it lands far below, some distance from the mountain's base. He imagined that the person keeps this up, throwing harder and harder, and the stones land not only farther and farther away but down around the side of Earth. After a while one is thrown so hard it sails around Earth and approaches the person from the back side, to continue circling endlessly. This circling stone is always falling toward Earth, but the ground is receding from it at a rate that matches the stone's rate of fall. This is true irrespective of the size or weight of the stone. So it is true of that great stone, the moon.

Newton did not stop here. He imagined the person still throwing stones, harder and harder. At some point in his thought experiment, stones were thrown beyond Earth's power to recall them. Newton had deduced the idea of the construct called "escape velocity."

Res. Ex. 7.25 Are there two horses on Earth with the same number of body hairs?

It's impractical to know by counting the hairs on all the horses. But what a work saver is a thought experiment. Thirty seconds of thinking will reveal the fact.

Imagine that 100,000,000 horses are alive at present, a ballpark figure good enough for a reliable thought experiment. Imagine that 50,000,000 hairs are the most that a horse can have (certainly the hairiest horse in history had fewer than that). Imagine we have a line of 50,000,000 stalls, numbered 1, 2, 3, . . . , up to 50,000,000. Imagine we errorlessly count the number of hairs on each of the 100,000,000 horses, and put each horse in the stall whose number is the number of hairs on the horse. Clearly, there must be at least one stall with at least two horses in it, proving there are at least two horses with the same number of hairs. As well, it proves there is a high probability that there are many pairs, triplets, etc., of horses having the same number of hairs. (Thanks to Michael Starbird for this example. He calls its form of deduction "the pigeonhole principle." Whenever there are more entities than pigeonholes to file them in, after filing there must be at least one pigeonhole with more than one entity in it.)

Res. Ex. 7.26: Is there a duplicate of our universe?

Some cosmologists believe that our universe is one of an infinite number of universes. Ours and these other universes are finite, and so there is less than an infinity of possible configurations of the matter and energy in them. Thus by the pigeonhole principle, there must be an infinite number of copies of our universe (and each of the others), down to the last detail, including you reading this book at this moment (Seife 2004). Moreover, there are an infinite number of copies that just miss being our universe, some of which have you building Microsoft from scratch while Bill Gates is a pauper. Naturally this thought experiment is premised on the existence of an infinite number of universes. Provided the premise is true, by deduction it follows that identical universes exist.

Res. Ex. 7.27: Steel cubes, granite mountains, and basketball leagues

Put on your imagining cap: (1) Imagine you are sitting on a piece of steel, 2 feet long, 2 feet wide, 2 feet high. Imagine it starts growing at the same rate in all directions. Its volume and therefore its weight increase with the cube of its height, while the area of its bottom face increases with the square of its height. Accordingly, the pressure on its bottom face builds disproportionally. When it is nearing four miles high, call for the rescue helicopter. The largest possible steel cube, it can be deduced, is about four miles high. Beyond that, its bottom crumbles under the pressure of its weight (Garfunkel 1987).

(2) Imagine a granite molehill you are standing on starts growing. Better climb down before it gets about seven miles high. It can be deduced that the tallest possible granite mountain is about a mile higher than Mt. Everest. Any higher and the coherence of granite gives out (Garfunkel 1987).

As a rule, with thought experiments involving facts of the strengths of materials, we can deduce the maximum possible size of anything.

(3) It's a good bet that basketball leagues will never have to raise the official height of the rim to compensate for taller players than exist now. Because, humans are close to the limit of height for a creature able to walk upright on two legs. How can that be known? Went (1968) imagined a 2-meter tall person, and from that deduced that the person when tripping "will have a kinetic energy upon hitting the ground 20 – 100 times greater than a small child who learns to walk. This explains why it is safe for a child to learn to walk; whereas adults occasionally break a bone when

tripping, children never do. If a man were twice as tall as he is now, his kinetic energy in falling would be so great (32 times more than at normal size) that it would not be safe for him to walk upright. Consequently man is the tallest creature which could reasonably walk upright on two legs."

Res. Ex. 7.28: What would happen if all of Earth's negative charges were stripped away?

It's not a pretty thought experiment. In a split nanosecond, you would be crushed into the ground by the net attraction of Earth's positive charges for your negative charges. What if instead Earth's positive charges were stripped away, leaving its negative charges? Instantaneously you would be repulsed into space at unbelievable speed. Listen to Ford (2004): "[The electric force between a hydrogen atom's proton and electron] outpulls the gravitational force by a factor that can truly be called humongous: more than 10^{39}. (How big is 10^{39}? That many atoms, stacked end to end, would stretch to the edge of the universe and back a thousand times.)"

Res. Ex. 7.29: Is the Sun a big ball of burning coal?

It's long been known that the Sun has burned for hundreds of thousands of years (if not, why do those fossil fish have eyes?). What is its fuel? Coal, some used to guess. In time, physicists did a thought experiment that proved otherwise. They began by imagining that the Sun was indeed a big ball of burning coal. They deduced how long it could burn, given its size and the energy coal contains. The imagined Sun quickly burned out, proving that coal cannot be its source of heat. Hoyle (1950) explains: "If the Sun were made out of a mixture of oxygen and the best quality coal, the coal would be reduced entirely to ashes in only two or three thousand years."

Res. Ex. 7.30: Did the movie *Jurassic Park* play loose with the truth?

Jurassic Park has *T. rex* running about 45 miles per hour chasing a jeep. This is the input to two thought experiments, described by Stokstad (2002):

(1) R. McNeill Alexander, from imagining a *T. rex* racing about 45 m.p.h., concluded that the leg bones of an adult *T. rex* would have cracked under the pounding forces of running.

(2) J. Farlow, from imagining a *T. rex* racing about 45 m.p.h., concluded that at times it would have tripped, a six-ton crash. Even with hospitals in those days, *T. rex* would have been D. O. A.

And so, the imagined existence of speedy *T. rex*'s leads to their demise, unable to pass on their speed genes, leaving breeding to slow moving *T. rex*'s. This must, as well, have been true in the real world, as no part of the deduction in the imagined world is dependent on which world it takes place in, imaginary or real.

Res. Ex. 7.31: How do we know that an ant can't build a fire?

We'll never know the answer from a real experiment. There is no way to train an ant to try to build a fire. Hence we imagine an ant determined to build a fire. It turns out that it dies trying. Went (1968) deduced that the size of the minimal sustainable flame is too large to allow an ant or even a small rodent to approach close enough to add the fuel necessary to keep the fire going. See Vogel (1988, p. 4) for comments on Went's thought experiment.

Res. Ex. 7.32: Does Santa Claus visit all the children on Christmas Eve?

I found the thought experiment that answers this question anonymously attributed and posted on the Internet under the title "SANTA CLAUS: An Engineer's Perspective." It includes several lines of deduction that all land on the same conclusion about Santa. I quote it.

"There are approximately two billion children (persons under 18) in the world. However, since Santa does not visit children of Muslim, Hindu, Jewish or Buddhist religions, this reduces the workload for Christmas night to 19% of the total, or 378 million (according to the Population Reference Bureau). At an average (census) rate of 3.5 children per household, that comes to 108 million homes, presuming that there is at least one good child in each.

"Santa has about 31 hours of Christmas to work with, thanks to the different time zones and the rotation of the earth, assuming he travels east to west (which seems logical). This works out to 967.7 visits per second. This is to say that for each Christian household with a good child, Santa has around 1/1000th of a second to park the sleigh, hop out, jump down the chimney, fill the stockings, distribute the remaining presents under the tree, eat whatever snacks have been left for him, get back up the chimney, jump into the sleigh and get on to the next house.

"Assuming that each of these 108 million stops is evenly distributed around the earth (which, of course, we know to be false, but will accept for the purposes of our calculations), we are now talking about 0.78 miles per household; a total trip of 75.5 million miles, not counting bathroom stops or breaks. This means Santa's sleigh is moving at 650 miles per second – 3,000 times the speed of sound. For purposes of comparison, the fastest man-made vehicle, the Ulysses space probe, moves at a poky 27.4 miles per second, and a conventional reindeer can run (at best) 15 miles per hour.

"The payload of the sleigh adds another interesting element. Assuming that each child gets nothing more than a medium sized Lego set (two pounds), the sleigh is carrying over 500 thousand tons, not counting Santa himself. On land, a conventional reindeer can pull no more than 300 pounds. Even granting that the 'flying' reindeer could pull ten times the normal amount, the job can't be done with eight or even nine of them – Santa would need 360,000 of them. This increases the payload, not counting the weight of the sleigh, another 54,000 tons, or roughly seven times the weight of the Queen Elizabeth.

"Six hundred thousand tons traveling at 650 miles per second creates enormous air resistance – this would heat up the reindeer in the same fashion as a spacecraft re-entering the earth's atmosphere. The lead pair of reindeer would absorb 14.3 quintillion joules of energy per second each. In short, they would burst into flames almost instantaneously, exposing the reindeer behind them and creating deafening sonic booms in their wake.

"The entire reindeer team would be vaporized within 4.26 thousandths of a second, or right about the time Santa reached the fifth house on his trip. Not that it matters, however, since Santa, as a result of accelerating from a dead stop to 650 m.p.s. in .001 seconds, would be subjected to centrifugal forces of 17,500 g's. A 250 pound Santa (which seems ludicrously slim) would be pinned to the back of the sleigh by 4,315,015 pounds of force, instantly crushing his bones and organs and reducing him to a quivering blob of pink goo.

"Therefore, if Santa did exist, he's dead now."

Res. Ex. 7.33: Could the universe be teeming with life?

Let's imagine a recipe for making the universe. We start with infinite empty space. Throughout it we sprinkle stars everywhere, all with one or more planets with life, making the universe teem with life. Going on, for each of these stars we sprinkle millions of decoys – stars without planets, and stars with planets not containing life. Everywhere we gaze out into this imaginary universe teeming with life, it looks like there is no life – just as it looks now. So, yes, the real universe could be teeming with life.

From thought experiments we can sometimes learn how to make something happen:

Res. Ex. 7.34: Could a baseball bat be made to fly?

I know, attach a parachute to it. Or treat it as a fuselage and add wings. Or drop it into a tornado. But there is another way to make things fly. To get perfume to fly, atomize it. To get tobacco to fly, convert it into smoke particles.

Here come spheres again! Imagine we have two sizes of spheres, each made of material whose density is one ounce per cubic inch (the actual density doesn't matter to the final result). The larger sphere's volume is 1,000 cubic inches, its weight 1,000 ounces. For it I calculated its diameter = 12.41 inches, its surface area = 484 square inches, its surface-to-volume ratio = 0.484, and its surface-to-weight ratio = 0.484.

The smaller sphere's volume is 0.01 cubic inches, its weight 0.01 ounces. For it I calculated its diameter = 0.267 inches, surface area = 0.224 square inches, surface-to-volume ratio = 22.4, and its surface-to-weight ratio = 22.4.

Pretend we drop the 1,000 oz sphere; it falls at the usual rate. Then pretend we reform it into 100,000 spheres, each weighing 0.01 oz. With that, each of the 0.01 oz. spheres has a surface-to-weight ratio that is $22.4 \div 0.484 = 46.3$ times larger than that of the 1,000 oz. sphere. This larger surface-to-weight ratio gives them greater air resistance. When dropped they will fall relatively slowly. Made even smaller, at some point a puff of air will waft them away.

For the baseball bat, increase its surface area relative to its weight by grinding it into individually invisible pieces. Find a breeze. The reformulated baseball bat will fly. In the same vein, imagine a pollen grain the size of a baseball. It will never fly. As Huxley (1929) explains, actual pollen grains, with their considerable surface area relative to their tiny weight, can travel way beyond their parent plants, the better for colonizing new areas. It is unnecessary that the ground-up particles of the baseball bat be exactly spheres, and that pollen grains be exactly spheres. Any shape, provided it is small enough, will fly.

Res. Ex. 7.35: How to make Earth survive a hit from a Manhattan-size asteroid

Imagine that some of Earth's people pray successfully that the jolt of the asteroid's hitting the upper atmosphere breaks it into chunks much smaller than city blocks. The volume of all the separate chunks will equal the volume of the Manhattan-size asteroid. But the sum of the surface areas of the separate chunks will be bigger than the surface area of the Manhattan-size asteroid. Being in chunks will make the asteroid vaporize in the atmosphere, leaving nothing to hit Earth's surface. Or imagine that some of Earth's people believe more in themselves than in the power of prayer. Imagine they send a rocket up to blow it to smithereens. Smithereens excel in large surface-to-volume ratios.

Res. Ex. 7.36: How to tell when a black hole lurks nearby

Astrophysicists John Archibald Wheeler and Bohdan Paczńyski (see Irion 2002) imagined that a black hole drifts past Earth, a dozen or so light-years away. And they imagined a flashlight shining on the black hole, a strong one, the Sun. They deductively reasoned the consequences of the Sun's rays approaching the black hole. The black hole will eat those rays. The rays just off the black hole's side will skirt the black hole, and its super-strong gravity will sling them around it. Some of the slung rays will make it for half an orbit, then escape and head back toward the Sun and us. This will make a ring of light around the black hole. Some of the other rays will be slung around for one and a half orbits, forming a smaller ring of light, closer to the black hole. With a large telescope, astronomers should be able to see the concentric rings. And so by this thought experiment, should someday such rings be spotted from Earth, it would be reasonable to hypothesize that a black hole was at their center.

The message of this chapter is that a reasoning model – a deductive method of reasoning – is a scientific method. There are three kind of reasoning models, making three kinds of scientific methods. With one, we deduce a fact – predict it – from one or more known facts. With another, we deduce a fact not because we want to know what the fact is, but because we want to explain the process that connects the one or more known facts to the predicted fact. With the third, we do thought experiments. We assume a fact or facts, and from them we deduce a fact to discover the consequence of our assumption.

Of all knowledge, among the most reliable is that produced with reasoning models.

8

How to discover knowledge by building and using computer simulation models

Like reasoning models, computer simulation models deductively predict facts. But unlike reasoning models, which predict qualitative or quantitative facts, computer simulation models predict quantitative facts. And unlike reasoning models, which are a few steps of deductive logic, and tend to be built by one person, computer simulation models are none of this. They are hundreds of steps of deductive logic, with a great many variables in interlinked equations, and almost always are built by the combined labors of two, three, or more researchers; usually they are too big of an intellectual job for one to do.

An example: A team of forest hydrologists decides to build a computer simulation model to predict the daily amount of runoff from a watershed's melting snow. They make a network of equations that represents the relevant physical processes. They translate the equations into a computer program. They read into the program input facts: depths of snow, air temperature readings, cloud cover percentages, and so on. They run the program. At locations in the watershed, it predicts the daily amounts of water running off, of groundwater storage, of evapotranspiration, and more. And provided the researchers have reason to believe that the computer simulation model is reliable, they may run it with hypothetical inputs of extreme snowpack depths melted in record time, to see how the model watershed handles the surge.

The facts that computer simulation models predict may be about what once happened, about what is happening, or about what will happen. Cosmologists build and use computer simulation models to predict aspects of past, present, and future star formation.

Our trip through this chapter begins with a way of thinking about computer simulation models (section 8.1). Then we cover three uses of computer simulation models that parallel the three uses of reasoning models in chapter 7. They are simulation models for predicting facts (section 8.2), simulation models for explaining processes of nature (section 8.3), and simulation models for doing thought experiments (section 8.4). The trip ends with a list of ten rules for building computer simulation models that will be as reliable as can be (section 8.5).

8.1 A way of thinking about computer simulation models

Let's pretend that aliens land their spaceship in a junkyard. They find a car's engine. The big picture of how it works is a mystery to them. So they tear it down. They study each part, what it does and how it is attached to nearby parts. By this they get little-picture knowledge of how the pistons work with the connecting rods, little-picture knowledge of how the timing belt works with the crankshaft, little-picture knowledge of how the battery works with the electronic distributor, little-picture knowledge of how the fuel injectors work with the fuel pump, and so on. All the same, a complete set of little-picture knowledge gives them no reliable idea of the big-picture knowledge of how the engine would work if they could actually run it. What the parts do is clear; what the engine does, the system assembled of the parts, is a mystery.

But it's not mysterious for long. They build a computer simulation model of the engine. They make it out of mathematical variables standing for aspects of the engine's parts, with the variables linked in equations. To give the model life, movement, they recast the variables and equations as a computer program. With a push of the computer's Go button, they run the simulation model, the stand-in for the real engine. This gives them knowledge of the big picture, the system of interacting parts. And it lets them predict facts, such as the maximum torque the engine can produce.

Researchers are in the position of these aliens. A certain engine of nature interests them. They know its little pictures, its conceptual parts and the parts they are attached to. They represent the parts mathematically, connected in equations. They translate the equations into a computer program. They run the program to discover big-picture knowledge of how nature works, and to predict facts of nature. If they know their business, which means they have made the model satisfy the ten rules of section 8.5, they can expect it to predict accurately, be reliable.

The variables and parameters in computer simulation models. There are three kinds of variables in simulation models: input variables, output variables, and intervening variables.

Input variables are the variables that we supply the numerical values of. Ordinarily, the values are from data, i.e., estimated facts of nature. (An exception: When the model is used in a thought experiment, we assign hypothetical values to one or more of the input variables, as explained in section 8.4.) Another name for the values of the model's input variables is the model's "initial conditions."

Output variables are the variables whose values we want to know by having the model predict them. The material in this chapter applies to models with one or more output variables. But to keep the discussion tight, we will mostly discuss models with one output variable. We will call its value "the model's prediction" or "the predicted fact."

Intervening variables are variables within the model's equations, between the input variables and the output variable. When we run the model, the values of the intervening variables are computed. These values may help us understand the model's behavior and judge its reliability.

As to **parameters**, they are numerical values in the model's equations. Usually the parameters in a model do not change when the model is run with different values of its input variables. Think of an epidemiological model of the spread of rubella in schools. It has a parameter whose value is the probability of transfer of rubella from one student to another. If the model is run for particular schools, the value of the parameter is kept the same, since it is independent of the particular schools. In some cases, parameters have been estimated and published in reports or handbooks. In other cases, they haven't been published, so must be estimated.

Deterministic and probabilistic computer simulation models. A deterministic computer simulation model has its input variables set at their mean (average) values. A single run of the model produces the mean value of the output variable. A probabilistic computer simulation model (also called a stochastic model) has its input variables set as random variables, their values from probability distributions. A probabilistic model is run a great many times, according to what is called the Monte Carlo method. For a run, the computer selects at random the values of the input variables from their probability distributions. The runs produce probability distributions of values of the intermediary variables and of the output variable, as well as the means and variances of the distributions. What we will say in this chapter holds for both kinds of computer simulation models. But because deterministic models are simpler to explain, we'll focus on them.

8.2 Using computer simulation models to predict facts

From one or more facts we know, suppose we want to predict a fact we don't know. Why not just measure it? Measuring it could be prohibitively expensive. Or it could be about the past, such as how many eclipses of the Sun were visible from Iceland during the sixth century. Or it could be about the future, like the number of feet above or below flood stage that the Ohio River will crest at Cincinnati next week.

Computer simulation models to predict facts are used in every field of science and commerce. There are simulation models that predict the spread of oil from spills at sea; simulation models that predict the spread of fire through certain chaparral ecosystems; simulation models that predict the seepage of toxic wastes into aquifers; simulation models that predict the growth in demand for social services, as for medical care and education; and simulation models that predict (forecast) the available pool of U. S. Social Security funds at the time today's babies retire.

The P-plane diagram for a computer simulation model resembles a cantilevered bridge. Measured facts support it at the model's input end, and from there its linked equations in the C-field span over to the fact that the model computes, predicts. The diagram is like the simple one for reasoning models (Fig. 7.1), except the network of linkages in the C-field is a hundred or more times as intricately involved.

Res. Ex. 8.1: Forewarning about the possibility of "lights out"

Perish the thought of an asteroid the size of Manhattan Island slamming into Earth. Former astronaut Edward Lu has remarked, "This is not just getting hit and killed. You're on the other side of the Earth and the atmosphere turns 500° hotter. Lights out." So, embrace the thought of predicting with a simulation model, years in advance, the fact that an asteroid is on a collision course with Earth, allowing time to send a rocket to nudge it into a harmless path.

Stone (2008) describes a simulation model that predicts the probability of any of the known asteroids hitting Earth during the next thirty years. Into the model go reliably estimated facts, including the given asteroid's and Earth's positions, masses, speeds, and trajectories. After calculations, out of the model comes the estimated probability of a collision. To date, for all known asteroids the probabilities are nil. Yet there will be the day when for an asteroid the probability isn't nil, and the call will go out to the rocket experts.

The equations in asteroid simulation models are built on Newton's laws. The facts of the input variables and parameters are precisely measured and contain negligible systematic error. In the language of section 8.1, the models incorporate highly reliable little-picture knowledge, and consequently they deliver highly reliable big-picture knowledge.

Res. Ex. 8.2: How many deer do we expect in three years?

Mills (2007) offers a good account of how wildlife ecologists build and run computer simulation models known as "population-projection models." The models predict the number of animals – ungulates, snakes, frogs, whales, or such that will be alive at a future time. I'll sketch how the practice goes for a deer herd. The model's input variables are the estimated number of deer at present, by sex and age. The model's equations and parameters represent the deer's birth and death processes from year to year. Among the model's intervening variables are the numbers of deer by age and sex at times between the present and the time we want the prediction for, suppose three years hence.

The parts of the model amount to inexact little-picture biological knowledge (inexact relative to exactness in physics). The inexactness is partly from random and systematic errors in the estimates of the values of the input variables and parameters. And it is partly from systematic error in the equations, also called "specification error," due to the model imperfectly specifying the birth and death processes. As a result, the model's prediction is inexact. Whether it is exact enough to help wildlife managers with their decision making depends on the species and its situation, as well as on other surrounding information the managers have.

8.3 Using computer simulation models to explain processes of nature

Our aim here is to explain (meaning to understand, to get insight into) the process that converts the facts that we know – the input to the simulation model – to the fact that the model predicts. Running the model reveals big-picture knowledge of the process, which counts as the process' explanation.

Think of research in astrophysics, where computer simulation models are built and run to get insights into processes like those of sunspot activity. And think of traffic research, where computer simulation models are built and run to learn about traffic flow. A simulation model I'm familiar with predicted the traffic dynamics as vehicles slow to a crawl past the site of an accident, and later try to resume normal speed. By studying the big-picture workings of the model, the traffic modelers explained how waves of stop-and-go driving form and evolve, rippling down the freeway.

Res. Ex. 8.3: How do engine lubricant additives work?

Trial and error has been the usual way of discovering new additives that do a better job of forming protective films on rubbing surfaces, reducing wear. Listen to Mark Robbins, a friction specialist at John Hopkins University (Goho 2005): "People traditionally tried a whole bunch of lubricants and found that some worked and some didn't, but it wasn't always clear why. Without understanding why something works, it's very hard to figure out how you would select a different lubricant that might address a slightly different problem."

Mosey et al. (2005) built a computer simulation model whose parts were absolutely reliable knowledge of quantum chemical processes of a zinc dialkyldithiophosphates (ZDDPs) additive. When run, the model revealed the values of its intervening variables and its output variables. This provided insights at the molecular level of how antiwear films protect metal and dissipate energy. In their research article, Mosey et al. say, "Here we report quantum chemical simulations of tribochemical reactions under extreme nonequilibrium conditions. The results of these simulations suggest that the formation, functionality, and frictional properties of ZDDP antiwear films are due to pressure-induced changes in the bonding at Zn atoms that transform an initially viscoelastic system of zinc phosphate chains into a chemically cross-linked network. The degree of connectivity within the system, and the resulting elastic properties, are found to depend sensitively on the maximum pressure to which the system has been exposed, in a manner that correlates well with the known properties of ZDDPs on steel surfaces. Because atoms with flexible coordination numbers other than Zn can act as cross-linking agents, we suggest that this mechanism can be extended to other systems. For example, similar processes may occur in artificial joints, where it is observed that proteins decompose and form interconnected carbon-based sheets similar to graphite that prevent wear." They go on to cite a diverse body of experimental data that their computer simulation model explains.

This is reliable science. The big-picture explanation of how engine additives work holds because the computer simulation model was built of errorless little-picture knowledge of quantum chemical processes. Moreover, it is great science. (1) The researchers discovered a cause-effect mechanism: that pressure is all that is needed to make the slippery films form, and not, as some researchers had surmised, temperature or chemical interactions between the additive and the metal's surface. (2) Their discovery explains previously unexplained experimental data. (3) Their discovery suggests extensions to other mechanical systems.

8.4 Using computer simulation models to do thought experiments

For a moment, refresh your memory of what a thought experiment is. I wrote in chapter 7: "Thought experiments are experiments in an imaginary world. They amount to posing a hypothetical, like 'What if this certain imagined view of nature is true?', or 'Assume that such is the case.' To run thought experiments, we deductively reason out their consequences. Sometimes their consequences reveal knowledge of how nature is. Other times they reveal knowledge of how nature is not. Either way, we learn something of nature." We went on in chapter 7 to show how reasoning models can be used to do thought experiments.

For our discussion in this chapter, let us re-express this definition of a thought experiment. It is that whenever we assign a hypothetical value to an input variable or a parameter, the simulation model is being used as a thought experiment. Let's have an illustration.

Res. Ex. 8.4: Which asteroids that are fated to strike Earth should we worry about?

To answer the question, Chyba (1993) devised a computer simulation model. With "what if" intent, he made a set of runs based on systematically imagined input facts, covering ranges of asteroid size, speed, entry angle, and material composition. For each run, the model predicted the fate of the hypothetical asteroid. Here is a sample of Chyba's conclusions: Stony or carbonaceous asteroids, smaller than 50 meters, traveling at a typical 50,000 kilometers per hour, plowing into Earth's atmosphere, will in all probability, no matter the entry angle, break into small pieces (increasing the surface-to-volume ratio) and explode harmlessly before they get to within 40 kilometers of Earth's surface. Even for a 50-meter stony or carbonaceous asteroid, heading straight in, traveling slowly, no significant damage is expected. (Slower moving ones experience lower atmospheric forces, making them less liable to break apart and burn up, and so they penetrate the atmosphere further.)

For iron asteroids, which are 6 percent of all, the model predicted severer consequences. At every size, speed, and entry angle, iron asteroids are more likely to strike Earth's surface. And for those actually striking, the model predicted the dimensions of craters they would make.

"Parametric analysis" is what engineers call this sort of what-if modeling. Engineers systematically vary the values of a model's input variables (treating them as parameters) through a range of settings, and for each setting they record the model's predicted fact.

Res. Ex. 8.5: Did the movie *Jurassic Park* play loose with the truth? – the sequel

Recall from Res. Ex. 7.30 that two thought experiments made with reasoning models strongly suggested that *T. rex* was incapable of running at the speed the movie has it running, about 45 mph. One indicated that the pounding of running would have fractured *T. rex*'s leg bones, and one that *T. rex* would have tripped often enough, killing itself, to go extinct.

Let's see how with a computer simulation model Hutchinson and Garcia (2002) reached the same conclusion. Based on the biomechanics of *T. rex*, they constructed a simulation model of an adult *T. rex* running, and used it in a series of thought experiments. They inputted the estimated weight of an adult *T. rex*, six tons. They imagined *T. rex* ran at various speeds from slow to fast. For each imagined speed, they ran the model and it calculated the power *T. rex* would have needed to move its six tons at that speed. And it calculated the minimum weight its leg muscles would have been to produce that power. At an imagined speed of 45 mph, its leg muscles would have been 86 percent of its entire body weight. This is impossible; its skeleton rules out its being 86 percent legs.

Biewener (2002) commented: "A pleasing aspect of Hutchinson and Garcia's study is that they apply sensitivity analysis [to their model] – allowing for a degree of parameter uncertainty – to evaluate the robustness of their results. . . . [Collectively the] uncertainties contribute to up to a threefold variation in the estimated muscle mass for the various models of *Tyrannosaurus*, but the conclusion is unchanged." (Part of coming section 8.5 shows how sensitivity analyses go.)

Stokstad (2002) pointed out the importance of the research, saying: "What does speed matter? Once an upper limit is established, Don Henderson of the University of Calgary in Alberta notes, paleontologists can put a cap on ecological questions such as how much territory a tyrannosaur could patrol in a day and how many top carnivores an area could support."

One more example of a thought experiment:

Res. Ex. 8.6: Were pterosaurs skim feeders?

Pterosaur is a flying reptile belonging to the genus *pteranodon*. (That pteranodon lived on cliffs along oceans is the conclusion of Res. Ex. 7.4's use of a reasoning model.) From its fossil remains, we know that a mid-sized pterosaur had approximately a 17 ft wingspan, weighed about 22 lb, and had an almost 3 ft long bill. Until recently, one school of researchers believed that pterosaur's elongated jaws suggest it was a skim feeder, like those present-day small birds that feed by flying close to the surface of water, skimming fish or small crustaceans with their lower bill. Another school, focusing on pterosaur's overall morphological features, doubted it was a skim feeder.

Humphries et al. (2007) settled the question by building a computer simulation model and using it for a thought experiment. One of the model's parameters was the drag force that pterosaur would have had to overcome in skim feeding. To estimate it, they made several different mockups of the lower bill of pterosaur, spanning the range of its fossil forms. For each, they measured the drag that the bill produced from water moving past it at various speeds and immersion depths. They entered each drag estimate into the computer simulation model and did an imaginary experiment of a pterosaur trying to skim feed. The model predicted the power it would have needed. Pterosaur, it turned out, was nowhere near capable of producing the required power to plough the tip of its bill through the water while managing to stay aloft.

Humphries et al. (2007) performed a sensitivity analysis on the model, varying each estimate of drag by an amount that would cover the range of uncertainty in the estimate. Always the conclusion was the same. A skim-feeding pterosaur would have crashed.

8.5 How to make computer simulation models be as reliable as they can be

As I said at the start of this chapter, computer simulation models are normally built by teams of modelers. Yet for the sake of speaking directly to you, I'll assume in this section that this is one of the exceptions, and you are building a simulation model by yourself. And even if you never build one, the rules will help you make up your mind about the reliability of those that are presented in research articles and seminars.

I'm going to give you ten sequential rules to follow to insure that the prediction the model makes, and the conclusion you draw from the prediction, are as reliable as possible. And although I will not discuss it, the ten rules also apply to building computer simulation models to explain processes of nature, and to do thought experiments.

Rules 1, 2, 3 and 4 apply to making preliminary judgments about the model before you build it. Rules 5, 6, 7 and 8 apply to assessing the reliability of the model. Rule 9 applies to drawing a conclusion from the model's prediction and assessing how likely it is that the conclusion is correct. Rule 10 applies to reporting on the model.

1 – MAKING PRELIMINARY JUDGMENTS BEFORE YOU BUILD THE SIMULATION MODEL (RULES 1, 2, 3 AND 4)

Rule 1: Never build a simulation model when any of the underlying natural processes being modeled are guesswork. Violate this rule, and the predictions from the model will in all likelihood be nonsense. I know of times when this has happened, and I will cite one. A computer simulation model of the life dynamics of the major animal species of a desert ecosystem was built and run despite the underlying natural processes being vaguely known (Payne 1974). The model consisted of several hundred equations. It had scores of input variables and parameters, and their estimated values contained large measurement errors. It amounted to an interconnected assembly of a great many little pictures of a desert ecosystem's parts, each more supposition than knowledge. When it was run, the expected happened: a humorously impossible view of the desert ecosystem resulted. While that was thirty years ago, not every modeler has learned the lesson. Witness the banks in 2008 that failed because their computer modelers included sheer guesswork in parts of the models they built for predicting the risks of lending money to consumers buying homes.

I have a little story about dinosaur skeletons that keeps me mindful of rule 1. Most of the typical dinosaur skeletons displayed in museums are made of plaster bones, as whole skeletons are rarely dug up. Listen to Mark Twain (Twain 1996): "Professor Osborn and I built the colossal skeleton brontosaur that stands fifty-seven feet long and sixteen feet high in the Natural History Museum, the awe and admiration of all the world, the stateliest skeleton that exists on the planet. We had nine bones, and we built the rest of him out of plaster of paris. . . . Nine bones and six hundred barrels of plaster of paris."

When a modeler guesses at the processes and the mathematical functions to represent them, the modeler acts the part of Professor Osborn. For those of us who are not modelers, we should be ready with a question at seminars. Ask in effect, "How much plaster is in your model, and how does that affect your conclusions?" The presenter is responsible for convincing us why the model is sufficiently faithful to the workings of nature. Remember, as little as 1 percent of a model that is guesswork – one or two plaster bones – can throw the whole thing into doubt.

Provided you adhere to rule 1, continue on.

Rule 2: Before you build the simulation model, state up front what you will conclude from it, depending on the value of the fact that it could predict. Draw this picture in your mind: You are the one who is building the computer simulation model in Res. Ex. 8.5, "Did the movie *Jurassic Park* play loose with the truth? – the sequel." Before you build it, you estimate from *T. rex*'s morphology the typical percentage of its weight that was in its legs. Say this is 15 percent. Then by rule 2 you state the following: When the model is built and run, if it predicts that an adult *T. rex* with 15 percent of its weight in its legs could have raced along at 45 mph, then conclude that the movie's portrayal is possibly true. Instead if it predicts it could not have, then conclude that the movie's portrayal is a lie.

Why, before you build the model, specify what you will conclude from it, depending on what it predicts? It's for the same reason that in sports we not permitted to switch the rules toward the end of close games.

Rule 3: Before you build the simulation model, decide how complex it should be. Modeling has its own version of the Goldilocks Principle. In building a given model, there is a just-right degree of complexity for it that minimizes the error in the fact that it predicts. Build a simpler model or a more complex model than that, and it will predict less accurately. Deciding on the just-right degree is an imprecise art, specific to each case of model building, but there are considerations for getting it approximately right.

To begin with, in building a simulation model to predict a fact, your aim is to minimize the error in the prediction. Alonzo (1968) breaks the **predictive error E** into its two component errors, e_m and e_s. The **measurement error e_m** is that part of E that comes from the measurement errors in the data for the model's input variables. The **specification error e_s** is that part of E that comes from the model's equations inexactly specifying the processes of nature being modeled.

Alonzo takes a model's complexity to be proportional to the number of calculations the model does to make its prediction. As complexity increases, meaning more calculations, a model's measurement error e_m compounds, increasing E, which is bad. At the same time, as complexity increases, a model's specification error e_s decreases, because the model better specifies the processes of nature, which is good. Fig. 8.1, adapted from Alonzo (1968), graphically shows the dependence of e_m, e_s, and E on the complexity of the model.

Alonzo (1968) writes: "The more complex the model, in the sense of having more operations [calculations] of the same kind or more 'explosive' operations such as raising to powers, the more the measurement errors cumulate as the data churn through their arithmetic. The gains in correctness of specification in a more complex model may be offset by the compounding of measurement errors. Although I lack the competence to demonstrate it, I am suggesting that if we tried to predict celestial phenomena by Einstein's General Theory of Relativity using data of the quality available to Copernicus, the predictions might be worse than if we used the Copernican theory."

Alonzo does not take into account that the more complex a model is, the more input variables it generally has, and this also acts to raise the measurement error e_m. Still, this doesn't change the aspect of Fig. 8.1 or change his message.

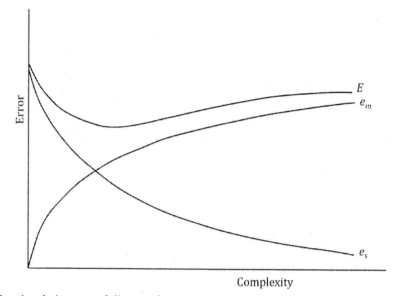

Figure 8.1. A simulation model's predictive error E and its two components errors, measurements error e_m and specification error e_s, as a function of the complexity of the model.

Rule 4: Have experts review your report describing the simulation model you are proposing to build.

Write a report for experts to review. In it, present your support for why the underlying processes of nature are sufficiently understood to model them. Describe and make flow charts of the model's conceptual parts and the connections among them. Write the model's equations. Lay out your plan for collecting data for the equations' input variables and parameters; tell how you intend to minimize the random errors in the data by setting proper sample sizes, and minimize systematic errors by using constructs that keep systematic errors low. Give an idea of the model's complexity by making a rough estimate of the number of calculations it will require. Which will it be closest to: 1,000 calculations, 20,000 calculations, 100,000 calculations, 2,000,000 calculations, or more? Say why its complexity strikes a balance between being too simple and being too complex.

Send the report to several experts (pay them well; you want their full-hearted attention). Request their reasoned opinions and suggestions for improvements. When they have finished their close examination, incorporate into the model's design any of their ideas that promise to improve its predictive accuracy. And should they advise you that too little is known of the processes to build a reliable enough model, drop the idea of going on unless you are able to convince yourself they are wrong.

2 – ASSESSING THE RELIABILITY OF THE MODEL (RULES 5, 6, 7 AND 8)

Assuming you are going on, translate the simulation model's equations into a computer program. Collect data and estimate the values of the input variables and parameters. Run the program (call this "the base case run") and learn what the simulation model predicts (call this "the base case prediction"). Then apply the next rule.

Rule 5: Check whether the simulation model behaves reasonably or unreasonably. For the base case run of the model, ask yourself the following. Is the value of the output variable, the model's prediction, possible in reality? Are the values of the intervening variables and (assuming a process over time is being modeled) their rates of change, and directions of change, possible in reality? A "No" to any of these questions means the model is wrong. In that event, check the model's equations and computer program for mistakes and oversights. Fix any you find, and re-run the model, revising the base case prediction. At the end, if the model cannot be fixed, don't go on with it. But if it behaves reasonably, continue to rule 6.

Rule 6: If feasible, test the simulation model with secondary data to learn how accurately it predicts. Primary data are the data you used to estimate the values of the input variables and parameters, and then ran the model with. But you cannot check how accurately the model predicts with its primary data, for you do not have the actual fact of the prediction.

Sometimes you can get secondary data and use them to see how accurately the model predicts. **Secondary data** are from processes of nature that are like the processes of nature that give the primary data. Provided secondary data are available, and provided they contain the actual fact that the simulation model is intended to predict, do the following. Run the model with the secondary data and judge how reasonably the model's intervening variables behave, and how accurately it predicts the actual fact. If it behaves reasonably and predicts accurately, that will raise confidence that it is correct. If it doesn't, that will raise the possibility that it is wrong.

An example: Suppose modelers build a simulation model of an airline's planned baggage facility. They want to predict the planned facility's response to bag jams, flights delays, and breakdowns of the conveyor line. They have access to secondary data from the airline's existing facility, which includes data on the frequencies of bag jams, flight delays, and breakdowns. They run the model with the secondary data. They check on how reasonable the values of the states and the rates of change of the intervening variables are. And they check how accurately the model predicts the response of the existing facility to bag jams, flight delays, and breakdowns. It might happen that the model's intervening variables take on unreasonable values, or that the model predicts poorly, raising doubts about it, leading the modelers to give it up. Or it might happen that the model passes the test with secondary data, boosting confidence in it.

Rule 7: Do a sensitivity analysis on the simulation model. A sensitivity analysis tells you how strongly dependent the model's prediction is on each of the input variables and parameters. The steps are the following. First, write down the value of the model's base case prediction that you got when you ran the model with primary data. Next, run the model a number of times. Each time, change the value of one input variable or parameter. Increase its value by a small amount, run the model, and record the value of the model's prediction. Then decrease its value by the same amount, run the model, and record the value of the model's prediction. Do this until all of the input variables and parameters have been changed, one at a time, up and down. This will tell you which of the input variables and parameters the prediction is highly sensitive to, i.e., which have large impact on the value of the prediction.

What size should the percentage increase or decrease in the value of an input variable or parameter be? It's rational to make the range – the increase in the value minus the decrease in the value – cover the probable error in the value. For talking purposes, in this section we will say this is about 10 percent.

There are three reasons for doing a sensitivity analysis. (1) It may give you insight into the model. (2) It will help you decide how confident you are that the model's prediction is correct. (3) It may show you how to make the model predict better. The three in turn:

1. Getting insights into the model: The sensitivity analysis identifies which of the input variables and parameters have great impact on the prediction, where small changes in their values produce large changes in the prediction. It reveals, so to speak, "the major and minor players" among the input variables and parameters.

2. Deciding how confident you are in the model's prediction. If the prediction is no more than moderately sensitive to changes in any of the values of the input variables and parameters, that is a plus. For that would maintain, perhaps even raise, your confidence that the model is correct. (Here, "moderately sensitive" is a subjective impression.) But if the prediction is highly sensitive (again, a subjective impression) to even a few of the changes, that is a minus. That would drop your confidence that the model is correct. How so? The values of the input variables and parameters contain measurement error. If the prediction is highly sensitive to one or more of the values, then it would be highly sensitive to measurement error. And a prediction that is highly sensitive to measurement error is a prediction that we cannot have great confidence in.

3. Making the model predict better: For those input variables and parameters causing greatest sensitivity on the prediction, it may be possible to collect more data to reduce the measurement errors in their values, and rerun the model with these data to get a more accurate, revised base case prediction.

As for reducing systematic error in the input values and parameters, whatever systematic error is in them is already minimized. You minimized it when you first collected the data.

Rule 8: State how likely it is that the model's prediction is correct. In reaching rule 8, you have a sense – a subjective probability of belief – of how likely it is that the model's prediction is correct. For thinking about this sense, and for conveying it to others, I suggest following Mahlman's (1997) recommendation. He advises modelers to subjectively estimate the certainty of their models' predictions in terms of the following three categories. *Virtually certain predictions*: predictions having a greater than 99 out of 100 chance of being true. *Very probable predictions*: predictions having a greater than 9 out of 10 chance of being true. *Probable predictions*: predictions having a greater than 2 out of 3 chance of being true.

Rule 9: Draw a conclusion from the model's prediction, and decide how likely it is that it is correct. There are three steps to this. (1) Return to rule 2 and note what conclusion you said you would draw from the model according to the value of its base case prediction. (2) Draw the entailed conclusion. (3) Decide how confident you are in this conclusion. To do this, review the sensitivity analysis you did under rule 7. There, for each run of the model, you changed an input variable or parameter, which produced a prediction. Examine these predictions now. If all of them lead you to draw the same conclusion that the base case prediction does, that will strengthen your confidence that this conclusion is the correct one to draw. But if the conclusion changes for one or more of the runs of the sensitivity analysis, that will weaken your confidence.

For this I suggest extending Mahlman's (1997) above recommendation to apply to conclusions modelers draw from their models' predictions. The extension has this form: *Virtually certain conclusions*: conclusions having a greater than 99 out of 100 chance of being correct. *Very probable conclusions*: conclusions having a greater than 9 out of 10 chance of being correct. *Probable conclusions*: conclusions having a greater than 2 out of 3 chance of being correct.

Rule 10: Thoroughly document the simulation model. Document the details of the model: its conceptual basis, the way its equations were devised to mimic processes of nature, how the values of its input variables and parameters were estimated, the prediction it made, the sensitivity analysis, the conclusion drawn, and such. So, do the necessary chore that modelers hate. Take a month or more to document the model to the extent that the whole project of building and running it could be duplicated by other modelers unfamiliar with the model. Have experts review the documentation for completeness and clarity.

An illustration of the rules. Let's assume you build and run a global warming simulation model according to the above rules. I want to discuss how you apply some of them. To begin with, at rule 1 you decide that enough is known of the processes of nature to build a model: You are pretty sure you can make the estimates of the values of the input variables and parameters be precise and contain a minimum of systematic error. And you believe you can make the error of specification in the model's equations be tolerably small. Another way to say this: you'd bet that the little-picture knowledge in your global warming model can and will be reliable enough to give acceptably reliable big-picture knowledge.

Meeting rule 1 allows you to go ahead to rule 2. There, you collaborate with decision makers who will decide whether or not to recommend enacting laws banning practices that could contribute to global warming. With their agreement you specify, say, that when you have the model built and run, that: (1) If it predicts a global temperature rise of less than or equal to $0.1°$ C per decade, it should be concluded that no laws are needed. (2) If it predicts a rise greater than 0.1° C per decade, it should be concluded that laws are needed. (To reach a final recommendation, the decision makers will weigh your conclusion along with those derived from other political and commercial information they have.)

When you run the model, it predicts, say, a temperature increase of 0.6° C per decade (its base case prediction). That is reasonably possible (an increase of, say, 3.0° C per decade is not reasonably possible). You also check the states and rates of the intervening variables. They are, let's say, reasonably possible: the global atmospheric convection acts normally, glaciers do not melt impossibly fast, and so forth.

You cannot apply rule 6 to test the model with secondary data. There is only one Sun and one Earth, and no data of global temperatures covering the last 100,000 years.

At rule 7, you perform a sensitivity analysis. By a small amount, say 5 percent, twice of which covers the probable error in the values of the model's input variables and parameters, you systematically change the values one at a time, up and then down, and run the model for each change.

All of the predictions in doing the sensitivity analysis are, say, within 0.15° C of the base case, a narrow band about it. Thus you decide it is unnecessary to collect more data to reduce the measurement errors in the values of the model's input variables or parameters.

At rule 9, you refer back to rule 2 where you specified: (1) If the model predicts a rise of less than 0.1° C per decade, it should be concluded that no laws are needed. (2) If it predicts a rise greater than 0.1° C per decade, it should be concluded that laws are needed.

So, the model predicts that the temperature will rise by 0.6° C per decade (the base case prediction). And throughout the sensitivity analysis there is no prediction of a rise of less than 0.45° C per decade. So, all predictions in the sensitivity analysis are well above a rise of 0.1° C per decade, making you confident to recommend that laws should be enacted.

Had it turned out the other way, with the model predicting, say, that global temperature would rise 0.12° C per decade, close to the critical value of 0.1° C per decade, the borderline where the conclusion changes, and/or had the sensitivity analysis bounced the prediction around in a range of, say, -0.05° C to +0.2° C per decade, this would leave you uncertain about concluding anything, be it enacting laws or not.

In your report (article or talk) you say that the global temperature will *probably* rise by 0.6° C per decade (i.e., you give the model's prediction a greater than 2 out of 3 chance of being right). And you say that it's *very probably* prudent to curb the apparent and known human-induced causes of global warming (i.e., you give it a greater than 99 out of 100 chance that taking this action is the correct conclusion.)

Notice that the certainty of the conclusion exceeds the certainty of the prediction that the conclusion rests on. How can that be? The conclusion is about the wisdom of not gambling with ruining Earth's biosphere, even though there is a moderate chance that the prediction is wrong, and global temperature will not rise by as much as 0.6° C per decade.

I have pushed several comments to the end.

Comment: As to the sensitivity analysis, I want to take note of why it is good practice to increase and then decrease the values of the input variables and parameters, rather than just increase them (as I have seen modelers do). The practice has two possible benefits:

1. Increasing and then decreasing each value will tell you whether the prediction changes in a nonlinear manner. If for the increase and then for the decrease, the predicted facts change from the base case prediction in opposite directions and by about the same absolute amount, this signifies a linear (proportional) response. But if for the increase and then for the decrease, the predicted facts change in opposite directions by different absolute amounts, this signifies a nonlinear (disproportional) response.

2. Increasing and then decreasing each value will help you better judge the reliability of the conclusion you draw from the model. To begin with, the base case prediction points to which conclusion to draw. But for the sensitivity analysis to make you confident that the conclusion pointed to by the base case prediction is the correct conclusion to draw, the sensitivity analysis must move the prediction through the full range resulting from increasing and decreasing the values of the input variables and parameters. To increase them only, or to decrease them only, doesn't give the full range of the prediction's "waggle," which may lead you to misjudge the reliability of the conclusion.

Comment: For simulation models commissioned by clients, Brewer and Shubik (1979) recommend three additional rules. An example of a client would be one that commissioned a traffic planning model for predicting the timing of traffic lights to minimize stop-and-go driving. The three rules: (1) Modelers should be sure they are addressing the problem their client really has, as opposed to the problem they "sort of think" the client has. (2) Modelers should be sure the client knows beforehand what the completed model will predict or explain. (3) Modelers should be sure the client understands the limitations of the model and the assumptions it rests on.

Comment: Consider rule 2, where you state what you will conclude from the simulation model depending on the value of the fact it could predict. The illustrations of the rule were the most common kind where the possible conclusions are two. Unillustrated were applications with more than two possible conclusions. For example, there could be applications with three (or more) possible conclusions. Each of the three (or more) non-overlapping ranges of the fact could entail a different conclusion.

Comment: The substance of a conclusion drawn from a simulation model differs according to whether the model is for basic research or applied research. Basic research is about gaining knowledge of what is – knowledge for the sake of just knowing. That *T. rex* could not run at 45 mph is a conclusion of the basic research simulation model described in Res. Ex. 8.5. Applied research is about gaining knowledge of what ought to be done – knowledge for the sake of acting on it. That we should act now to prevent global warming is the conclusion of the above applied research simulation model.

Comment: How do you perform a sensitivity analysis of a computer simulation model that has multiple output variables, i.e., predicts several facts? Perform one sensitivity analysis but keep track of its effects on each of the output variables, each of the predictors.

Comment: For a sensitivity analysis, there are two approaches to changing the values of the input variables and parameters:

1. Change one value at a time (as described above). For a model with, say, 8 input variables and 12 parameters, the values of each to be increased and decreased, there will be 40 runs of the model ($(8 \times 2) + (12 \times 2) = 16 + 24 = 40$).

2. Make all combinations of changes of input variables and parameters. This is practical only for models with few input variables and parameters; otherwise the number of runs will be beyond price.

Most of the sensitivity analyses I have seen change one input variable or parameter at a time. The approach is affordable, and it is easy to interpret the resulting changes in the intervening variables and in the predicted fact because they are caused by a single change in an input variable or parameter. However, the approach cannot indicate the sensitivity of the intervening variables and the predicted fact to interactions from simultaneous changes in the input variables and parameters, which making all combinations of changes can indicate.

I leave this chapter on a happy note. The future, I feel sure, will see computer simulation models play an expanding role in basic science and applied science. This is because simulation models are built of knowledge, and the prospects are for more and more little-picture knowledge, with it growing more and more reliable. Many of the limitations that now stop us at rule 1 – "Never build a simulation model when its underlying processes are guesswork" – will disappear. Who knows? – someday reliable computer simulation models of ecosystems may be possible. And someday near-term weather forecasts may be in the category of virtually certain predictions.

9
How to minimize the ethical biases in research

We have discussed how to minimize the methodological biases in research, e.g., bias in sampling. Now we will discuss how to minimize the ethical biases in research, e.g., bias when one researcher takes advantage of another. Although not every researcher is at risk of perpetrating ethical biases, enough do to invite discussion.

Before I give you my list of 21 ethical biases, I want to explain its source. From Owen's (1982) list of 25 kinds of ethical bias in medical research, I selected 15 that apply to all fields. To this selection I added four others from Jadad's (1998) list. And I added two that aren't on these lists but deserve to be (numbers 20 and 21 in my list). Although researchers may perpetrate the biases consciously, my guess is that most are done without realizing it. My list:

1. *Tit-for-tat bias.* Also called *I-owe-him/her-one bias.* Getting back at a researcher who found fault with your work. Ex.: Bob submitted a manuscript to a journal last year, and he knows that Joe reviewed it and recommended it be rejected. Now Bob has been asked to write a review of Joe's new book. . . .

2. *Prominent investigator bias.* Being disposed to uncritically accept results published by prominent researchers.

3. *Credential bias.* Downplaying the research of those without academic affiliation and/or a Ph.D.

4. *Mentor bias.* Giving a free pass to a research report or proposal by the professor who guided you in your graduate studies.

5. *Friendship bias.* Not rigorously searching for weaknesses in a friend's report or proposal.

6. *Dislike-of-person bias.* Having a bad opinion in advance of judging a report or proposal by a researcher whose personality has annoyed you, such as one who at a meeting seemed rude, quarrelsome, or pompous.

7. *Territory bias.* Disparaging a report or proposal in your field of expertise because its author is an outsider from another field.

8. *Language bias.* A predisposition to think more of research findings presented in one's own language than of those presented in a foreign language.

9. *Country of publication bias.* Trusting the reliability of research from highly developed countries more than that from lesser developed countries.

10. *Famous institution bias.* A report or proposal is by researchers at Harvard; ergo, unquestionably think it's reliable and substantial.

11. *Prestigious journal bias.* Overrating research just because it appears in a prestigious journal.

12. *Flashy title bias.* Self-explanatory.

13. *Personal value bias.* Showing favoritism to a report or proposal because it jibes with your personal values, or showing prejudice against one in opposition. Ex.: Professionally, Sam is a consulting marine scientist; personally, Sam likes to fish. A state's marine division hires Sam to evaluate a Scuba association's proposal to relax the rules limiting spearfishing. Sam is liable to make light of the impact the relaxed rules would have on creatures living in the state's coral reefs.

14. *Favored design bias.* Overrating research based on an experimental design you favor; underrating research based on one you disfavor. Ex.: A sociologist favors data-collection designs that have subjects answer questionnaires, and so sniffs at those using data from interviews.

15. *Latest statistical method bias.* Inclined toward research that relies on processing data with one of the latest statistical methods. Similar to this is *fashionable statistical method bias.*

16. *Avant-garde technology bias.* Uncritically believing in research that uses the latest technological thing.

17. *Empiricism bias.* Attaching little importance to research that deals in theory rather than data.

18. *Large sample size bias.* Holding that a research result is above reproach because it is based on a large sample of data.

19. *Resource allocation bias.* Depreciating a proposal because funding it will reduce the funds available in your area of research.

20. *Positive result bias.* Recommending that research that corroborates a hypothesis (a positive result) be published, but that research that disproves a hypothesis (a negative result) not be. Steven Wiley (2008) has written that "findings that disprove a hypothesis are largely not worth publishing" since they "don't actually advance science." I disagree. A published negative result is a stimulus to researchers to discover the error(s) in the support that initially justified testing the hypothesis. Such discoveries advance science.

Fifty years from now the above ethical biases could be things of the past. How? First, universities and research societies must educate researchers about the biases and how they threaten truth-seeking. For this to be effective, it will take instruction grounded with case studies. Second, outside the classroom, on their own, researchers must cultivate a commanding love of truth-seeking; then each will naturally be on guard against the enemies of truth-seeking, ethical biases.

We now come to the final ethical bias. Education and love won't cure it; it runs on the institutionalized quest for power and money.

21. *Kowtowing-to-the-one-that-funds-you bias.* The organizations that fund research – government agencies, companies, the military, private sponsors – are prone to suppressing discoveries of truths that could wreck their missions. As Jacques Cousteau put it, they "do not 'fund' science; they 'buy' science. Governments review the safety of defoliants, food additives,

leaded gasoline; they hire scientists paid by the industries under review and thereby purchase exactly the results most expedient for their plans. When tests could not be tilted so as to avoid a damaging truth, government agencies have sometimes contended that, since they have 'bought' the science, they 'own' the results and can refuse to release them. . . ." (Cousteau and Schiefelbein 2007, p. 198).

My friends, it takes a courageous ethical David to stand up to an unethical Goliath. The David's job is on the line; the Goliath's isn't. The only good option is for the Davids to band together. An instance of this is *FSEEE: Forest Service Employees for Environmental Ethics*. The Davids are U.S. Forest Service workers who care about discovering and disseminating environmental truths; the funding Goliath is the Forest Service, with some of its agendas served by skewing or silencing truths that would be politically and economically harmful to it if circulated.

Until each field's lone Davids merge to make a powerful force, they will be as helpless as a state university researcher I know. She worked in a research program that was supported with the state's farm agency funds. She proposed identifying areas in the state with habitats suitable for wolves. On learning of this, the farm agency, representing livestock interests, applied its clout. Unless the university got rid of the researcher, it would stop paying the university for research of any sort. The university acquiesced; it unceremoniously transferred her out of the program.

Fifty years from now, the Goliaths will still be with us. But by banding together, we can weaken their ability to suppress truth. *The Union of Concerned Scientists* and *FSEEE* are two of the bands that are showing the way.

10
Why the funding of basic research must be increased

Basic research isn't receiving from funding agencies the attention to which its importance entitles it. With a view to showing the folly of this, I have collected writings that explain its importance. As you read them, keep in mind the following equivalent terms. "Researcher" and "scientist" are the same. "Applied research," "applied science," and "practical science" are the same. "Basic research," "pure research," "basic science," and "pure science" are the same.

Jacques Cousteau defined applied research and basic research by their attractions to those who do them. A researcher who does **applied research** "longs to achieve a known objective" – "a prospector" who "sets out to find a 'useful' pot of gold." A researcher who does **basic research** "longs only to know the unknown" – "an explorer" who "sets out to find nothing" and "discovers the universe." (Cousteau and Schiefelbein 2007, p. 181)

Two sorts of social payoffs will come from boosting the funds for basic research. There will be material benefits (section 10.1), and there will be non-material benefits (section 10.2).

10.1 How basic research brings material benefits

Obviously, applied research leads to better goods and services: finer cars, cheaper airline flights, fuller energy utilization, new cures of disease, gains in conserving natural resources, and so forth. Yet increasing the funds for basic research will bring goods and services that surpass what applied research alone can bring. On this point, the following words of Ralph Gomory (Gomory 1983), Jacques Cousteau (Cousteau and Schiefelbein 2007, p. 192), and John Grier Hibben (Hibben 1908) need to be understood:

Hear Ralph Gomory:

"Science can be thought of as a large pool of knowledge, fed by a steady flow from the tap of basic research. Every now and then the water is dipped out and put to use, but one never knows which part of the water will be needed. This confuses the funding situation for basic science, because usually no specific piece of scientific work can be justified in advance; one cannot know which is going to be decisive. Yet history shows that keeping water flowing into the pool is a very worthwhile enterprise."

Hear Jacques Cousteau:

"Applied research generates improvements, not breakthroughs. Great scientific advances spring from pure research. Even scientists renowned for their 'useful' applied discoveries often achieved success only when they abandoned their ostensible applied-science goal and allowed their minds to soar – as when Alexander Fleming, 'just playing about,' refrained from throwing away green molds that had ruined his experiment, studied them, and discovered penicillin. Or when C. A. Clarke, a physician affiliated with the University of Liverpool, became intrigued in the 1950s by genetically created color patterns that emerged when he cross-bred butterflies as a hobby. His fascination led him – 'by the pleasant route of pursuing idle curiosity' – to the successful idea for preventing the sometimes fatal anemia that threatened babies born of a positive-Rhesus-factor father and a negative-Rhesus-factor mother."

Hear John Grier Hibben:

"Because this is a practical age, and also pre-eminently an age of extensive investigation, it might seem that the chief incentive to research would be the possibility of adding to the store of practical knowledge, and thereby increasing the general efficiency of human endeavor. But, as we read the history of scientific discovery, from the first strivings of primitive thought to the present time, we are impressed with the fact that utility is not always the mother of invention. This is the paradox which confronts us throughout the whole course of the development of scientific thought: If man interrogates nature for the purpose merely of wresting those secrets which shall minister directly to his needs or comfort, he fails to attain his end, or he attains it only in a meagre way; but if, on the contrary, he goes to nature with a desire to know her secrets for their own sake, the revelation often brings with it a wealth of knowledge which, in turn, admits of untold applications as regards the practical conveniences of life. If utility is the sole incentive to research, the results will range on a lower level; if, however, utility is forgotten in the passion to get at the heart of things for their own sake, it sometimes surprises us upon the way. And the reason of this is obvious; for utility, in all practical relations, results from the application of certain underlying principles to the concrete problems of life. The more central and comprehensive the principle, the wider will be its scope of practical application. The principles most fertile in products of utility are often most deeply hidden. They lie at the centre of things; it is only the most searching inquiry which will disclose them. . . . If you start in your research with the sole object of solving a specific problem of practical significance merely, the result, if successful, is limited in all probability to the special end in view; on the other hand, if you set yourself the larger problem of investigating certain phenomena which have peculiarly attracted your interest for the purpose of discovering their nature and understanding their laws, then the revelation of a comprehensive principle carries with it a whole world of possibilities. A principle is not one, but many, for it admits of a multiplicity of application which knows no limit. . . .

"A mind exclusively bent upon the idea of utility necessarily narrows the range of the imagination. For it is the imagination which pictures to the inner eye of the investigator the indefinitely extending sphere of the possible, – that region of hypothesis and explanation, of underlying cause and controlling law. The area of suggestion and experiment is thus pushed beyond the actual field of vision. But, if utility is the sole end of research, the scope of imaginative inquiry is thereby narrowed. There is no comprehensive sweep of the thought, no power of divination, no compelling fancy. Whatever fails to show a face value of utility does not arrest and hold the attention. Significant facts and relations are overlooked. The byways of knowledge are left unexplored in the hot pursuit of the immediately useful. But where there is absorbed and sustained interest in a subject of research for its own sake, the imagination broods over its tasks with a delight and passion which tend to provoke the hidden truth."

There we have it. Basic research is the means to material benefits that far surpass those that applied research alone is able to attain.

10.2 How basic research brings non-material benefits

W. W. Campbell (Campbell 1908) and Jacques Cousteau (Cousteau and Schiefelbein 2007, ch. 8) speak of basic research's non-material benefits:

Hear W. W. Campbell on how basic research influences thinking and feeling about the cosmos and our place in it:

"That the main results of the astronomer's work are not so immediately practical does not detract from their value. They are, I venture to think, the more to be prized on that account. Astronomy has profoundly influenced the thought of the race. In fact, it has been the keystone in the arch of the sciences under which we have marched out from the darkness of the fifteenth and preceding centuries to the comparative light of to-day.

"Who can estimate the value to civilization of the Copernican system of the sun and planets? A round earth, an earth not the centre of the universe, an earth obeying law, an earth developed by processes of evolution covering tens of millions of years, is incomparably grander than the earth which ante-Copernican imagination pictured.

"It is of priceless value to the human race to know that the sun will supply the needs of the earth, as to light and heat, for millions of years; that the stars are not lanterns hung out at night, but are suns like our own; and that numbers of them probably have planets revolving around them, perhaps in many cases with inhabitants adapted to the conditions existing there. In a sentence, the main purpose of the science is to learn the truth about the stellar universe; to increase human knowledge concerning our surroundings, and to widen the limits of intellectual life."

Hear Jacques Cousteau on how basic research inspires values:

"Science and poetry are, in fact, inseparable. By providing a vision of life, of Earth, of the universe in all its splendor, science does not challenge human values; it can inspire human values. It does not negate faith; it celebrates faith. When scientists discovered that liquid water, which brought forth life on Earth, exists nowhere else in great quantities in the solar system, the most significant lesson they taught was not that water, or the life that depends on it, is necessarily the result of some chemical accident in space; their most important revelation was that water is rare in infinity, that we should prize it, preserve it, conserve it."

Hear Jacques Cousteau on how basic research can lead people to protect nature. Our everyday view of the natural world is inadequate to awaken a sufficient protective love of the natural world. To awaken the sufficient love we need to see the natural world in many narrower and wider views. Applied science is blind to these views. Basic research is all eyes.

Cousteau:

"The reason I have made films about the undersea lies simply in my belief that people will protect what they love. Yet we love only what we know. Learning science, learning about nature, is more than the mere right of taxpayers; it is more than the mere responsibility of voters. It is the privilege of the human being."

Hear Jacques Cousteau on how basic research influences thinking and feeling about how all forms of life are closely related:

"[Basic research] has revealed to us that, in the infinite span of time and space, the similarities shared by all human beings carry far more significance than any differences between individuals. Of all the billions of genetic commands our bodies obey, 99 percent are identical to an ape's. Just 1 percent of our genetic makeup distinguishes us as human. The infinitesimal fraction of genetic material that distinguishes one human from another is negligible. The genes that distinguish a human who lives in one country from a human who lives in another simply do not exist."

Hear Jacques Cousteau on how basic research can bring peace. Our everyday view of the natural world is inadequate to end our unreasoning awe or fear of the unknown. Nor can applied research ever end it. Only basic research, by showing us all sides and depths of nature, can. Cousteau:

"Albert I of Monaco, a great seaman, an inspired oceanographer, and a generous patron of pure science, established the Oceanographic Museum in Monaco in 1910. Having searched his archives, having read his letters, now stored in palace vaults, I almost feel I know him. Albert believed that science could dispel the darkness of superstition, that it could bring peace to Earth."

The point is, I think, thoroughly made. Basic research influences us, inspires us, and educates us.

For the future we must break the existing order of things. One, the financial restrictions on basic research must be lifted. How? We need an army of persistently squeaky wheels. If researchers get involved, be vocal supporters for increased and indelible line-item funding of basic research, governments might be swayed.

Two, science professors must teach every science student the methods for reliably gaining basic knowledge. A trillion dollars could be made available for basic research, and yet the material and non-material benefits will not come if the research is done unreliably. Students today are not learning how to accurately evaluate the reliability of reports of basic research, nor how to maximize the reliability of basic research they might someday produce. Most do not know the effective ways of conceiving and reliably supporting research hypotheses. Most do not know the effective ways of reliably testing research hypotheses. Most would fail a test of elementary questions drawn from this book. That's the existing educational order, and our responsibility is to break it.

11

Tips on how to become a better researcher

Are you a student preparing for a career in basic research, or are you in the midst of your career? I have for you some tips for becoming a better researcher.

Take courses in probability and in statistics. Take an introductory course in probability theory and its applications. It will improve your skills of making decisions under risk, which in itself is a great plus. And it will benefit you later when you study inferential statistics: it will raise you above the plug-and-chug mentality that afflicts many students; you will grasp what inferential statistics is about. As for which course in probability theory to take, my advice is one with the coverage in Warren Weaver's book, *Lady Luck: The Theory of Probability.* Get Weaver's book and show it to a professor who offers courses in probability theory. Say, "I want to learn the topics in this book. What course do you recommend?"

Take courses in statistics. Geedey and Dudycha (2004) made a remark to ecologists that holds for researchers in every field: "If you can't understand enough statistics to interpret the data from your own experiments, then you probably don't deserve a Ph.D. in ecology." Start with the introductory course that covers descriptive and inferential statistics. At a minimum, follow it with courses in survey sampling, analysis of variance, regression analysis, and multivariate analysis.

Read books on scientific methods. Round out the material in this book by reading others on scientific methods. I am not going to suggest any; every reader reacts to a book differently. But I want to mention the kinds of books I have in mind. Here in no particular order are some I have read: *The Logic of Scientific Discovery*, by Karl Popper; *A Beginner's Guide to Scientific Method*, by Stephen S. Carey; *Scientific Method for Ecological Research*, by E. David Ford; *Theory of Scientific Method*, by William Whewell; *The Art of Scientific Investigation*, by W. I. B. Beveridge; *Scientific Methods in Practice*, by Hugh G. Gauch, Jr.; *Yes, We Have No Neutrons: an eye-opening tour through the twists and turns of bad science*, by A. K. Dewdney; *Chase, Chance, and Creativity: the lucky art of novelty*, by James H. Austin; *A Primer on Natural Resource Science*, by Fred S. Guthery.

Every few months spend an evening thinking up important research problems to solve. I call this the Alvarez method. Walter C. Alvarez practiced it, and he advised his son Luis to practice it. Hear his son, Nobel Prize winner Luis W. Alvarez (Alvarez 1987, p. 57):

"Dad never claimed he might have won that Nobel Prize, but he did say more than once that he would have been a better researcher if he had occasionally let his mind wander over the full range of his work. He advised me to sit every few months in my reading chair for an entire evening, close my eyes, and try to think of new problems to solve. I took his advice very seriously and have been glad ever since that I did."

Work on important research problems for which you have a secret weapon for solving them. I call this the Simon method. Among the important research problems you could work on, pick one for which you have a secret weapon, an edge based on superior preparedness: special knowledge or methods that put you on the inside track to solving it. Hear Nobel Prize winner Herbert A. Simon (Simon 1991, p. 111):

> "I advise my graduate students to pick a research problem that is important (so that it will matter if it is solved), but one for which they have a secret weapon that gives them some prospect of success. Why a secret weapon? Because if the problem is important, other researchers as intelligent as my students will be trying to solve it; my students are likely to come in first only by having access to some knowledge or research methods the others do not have."

*   *   *

There you have it. A book of best research practices based largely on my studying top scientific articles and organizing them according to the objectives of the research and the methods of achieving the objectives.

May it serve you well.

Appendix: Review of statistical terms

The main statistical terms appearing in chapter 2 are these:

Sample units are tangible things that can be sampled. As a running example, a sample unit is, say, an acre of land in Saguaro National Monument.

The **target population** is the set of sample units, N in number, that we want information about. N may be unlimited in number, or N may be finite in number. For the running example, the $N = 71,400$ individual acres comprising Saguaro National Monument is the target population. Statisticians and researchers often shorten "target population" to "population."

A **random sample** is a set of n sample units selected randomly from the N sample units of the target population. The symbol n is called the **sample size**. It might be that we randomly select $n = 84$ acres from the $N = 71,400$ acres in Saguaro National Monument. One way of doing this is to lay a grid of one-acre sample units on a map of the Monument and use random numbers to select 84 of the one-acre sample units.

On each sample unit in the sample we measure a variable of interest ("construct" is another name for "variable"; a "variable of interest" is a "construct of interest.") The symbols for the n measurements are $x_1, x_2, x_3, \ldots, x_n$. In the saguaro example, the variable of interest could be the number of saguaro cacti per acre, with $n = 84$ measurements of the number denoted by $x_1, x_2, x_3, \ldots, x_{84}$. A value of, say, $x_1 = 14$ would indicate that on the acre in the sample that we label with a "1" there are 14 saguaros, as measured by counting them.

The **sample mean**, denoted by $\overline{X}$, is the mean (i.e., the average) of the n measurements. For the saguaro example, we add the 84 sample facts $(x_1, x_2, x_3, \ldots, x_{84})$ and divide the sum by 84, the value of n, giving $\overline{X}$.

The **population mean**, denoted by the Greek letter μ, is the mean of all the values of the variable of interest in the population of N sample units. In place of knowing the actual value, μ, we accept a tolerably accurate estimate of μ, namely $\overline{X}$, along with an estimate of the reliability of the estimate.

The invention of the methods of statistical inference revolutionized research. In the early days when the methods were just coming into use, Walker (1939) wrote: "The idea the information obtained from a relatively small number of cases actually examined can be used to throw light on the characteristics of a vast universe which has not been examined is an exciting idea, which ceases to amaze us only when familiarity renders it commonplace. That the sample not only furnishes an estimate of some characteristic [a parameter, such as μ, or σ, the standard deviation] of the unknown population, but also furnishes a measure of the degree of confidence [a confidence interval] which can be placed in that estimate, is still more remarkable."

Chapter 2 gives more space to estimating confidence intervals about parameters of populations, e.g., the mean, μ, than it does to testing statistical hypotheses about parameters of populations. This is in keeping with researchers in most fields estimating parameters more often than they test statistical hypotheses about parameters; a look through journals such as *Science* and *Nature* confirms this. Just the same, chapter 2 does not completely gloss over statistical hypothesis testing. It is the thing to do when you want to compare the values of a parameter in two or more populations, as with an experiment involving a control group and an experiment involving a treatment group.

A **95-percent confidence interval** estimates the value of a parameter of a population, such as μ, and simultaneously estimates the degree of confidence in the estimate of the parameter. A 95-percent confidence interval is denoted by, and computed with, this equation: $\overline{X} \pm (s_{\overline{X}})(t)$, which written out is $\overline{X} - (s_{\overline{X}})(t) \leq \mu \leq \overline{X} + (s_{\overline{X}})(t)$. Here, $\overline{X}$, the sample mean, is an estimate of the population mean, μ; $s_{\overline{X}}$ is the sample standard error of the mean, which is an estimate of the population standard error of the mean, $\sigma_{\overline{X}}$; and t is the t-value, as tabulated in a t-table for 95-percent confidence. The phrase "the standard error of the mean" is commonly shortened to "the standard error." In practically all fields of research, 95-percent confidence intervals are the norm. At times, you will run across researchers computing "other-percent" confidence intervals, such as 99-percent or 90-percent. As for why 95-percent is the norm, see Romesburg (2004, p. 248–49).

Sample size determination refers to calculating the sample size *n*, the number of sample units to select randomly from a population of *N* sample units. The calculation is made with an equation. The researcher provides input information that includes the "percent confidence" wanted, and the "allowable error" wanted (the allowable error is how close the researcher would like, on average, $\overline{X}$ to be to μ). For the saguaro example, the sample size was (we are pretending) determined by calculation, giving a random sample of *n* = 84 sample units.

Now we turn to nonrandom samples. A **nonrandom sample** is a set of *n* sample units selected in any way except randomly. For instance, we could select *n* = 84 acres nonrandomly by numbering on a map each of the *N* = 71,400 acres in Saguaro National Monument and then choosing every 850th acre systematically (71,400 / 850 = 84). When nonrandom samples are used in research, the sample sizes are set not by an equation but by considerations of expense or circumstance. By expense: the rule is to make them as large as we can afford. By circumstance: in some areas of research, e.g., archaeology, the size of a nonrandom sample is set by the number of artifacts (the sample units) discovered. Ten stone tools are all that are dug up, making *n* = 10; each is weighed, giving ten measured weights, $x_1, x_2, x_3, \ldots, x_{10}$. The mean weight of the ten, $\overline{X}$, is computed. Valid confidence intervals cannot be obtained with nonrandom samples. Instead, researchers who work with nonrandom samples state in their research reports the sample mean, $\overline{X}$, and the sample standard error, $s_{\overline{X}}$, which gauges the variability within the set of sample measurements, $x_1, x_2, x_3, \ldots, x_n$.

(For more explanations of the above terms, see any book on survey sampling.)

A **standard error interval** is defined as the sample mean, plus or minus the sample standard error of the mean. Statisticians write it as $\overline{X} \pm s_{\overline{X}}$. Scientists sometimes write it as $\overline{X} \pm SE$, or as $\overline{X} \pm SEM$. Researchers take the standard error interval as indicating the variability within a sample of measurements. The narrower the standard error interval is, the less the variability is; the wider, the greater the variability is.

With a random sample, research protocol calls for computing a confidence interval. With a nonrandom sample, research protocol calls for computing a standard error interval.

A **statement of accuracy** is the researcher's estimate of the size of the systematic error in an experiment. A common form of a statement of accuracy reads: "the sample mean, $\overline{X}$, is believed accurate within $\pm$ i units." Think of this as an accuracy interval, i.e., $\overline{X} \pm i$. The value of i is subjectively determined by the researcher, according to professional judgment. A statement of accuracy tells what you believe is the residual size of the systematic error that remains after you thoroughly tried to minimize it. By analogy, if systematic error is water in a towel, you are telling readers how much water remains after you have done your best to wring it all out.

References

Alonzo, W. (1968). The quality of data and the choice and design of predictive models. Pages 178-192 *in* G. C. Hemmens, ed. *Urban Development Models.* Highway Research Board, Washington, D. C., Special Report 97. **Alvarez, L. W.** (1987). *Alvarez: adventures of a physicist.* New York: Basic Books. **Anderson, C. A.** and Bushman, B. J. (2002). The effects of media violence on society. *Science* 295: 2377-2379. **Andrade, C.** (2006). Transcendental meditation and components of the metabolic syndrome: methodological issues. *Archives of Internal Medicine* 166: 2553. **Andreae, M. O.**, Jones, C. D. and Cox, P. M. (2005). Strong present-day aerosol cooling implies a hot future. *Nature* 435: 1187-1190. **Angers, R. C.**, Browning, S. R., Seward, T. S., Sigurdson, C. J., Miller, M. W., Hoover, E. A. and Telling, G. C. (2006). Prions in skeletal muscles of deer with chronic wasting disease. *Science* 311: 1117. **Associated Press** (1995). Couple's cat finds its way across state. June 2, 1995. **Balph, D. F.** and Balph, M. H. (1983). On the psychology of watching birds: the problem of observer-expectancy bias. *The Auk* 100: 755-757. **Balter, M.** (2005). Are humans still evolving? *Science* 309: 234-237. **Barrett, J. P.** and Nutt, M. E. (1979). *Survey Sampling in the Environmental Sciences: a computer approach.* Wentworth, N.H.: COMPress. **Benson, M. A.** (1965). Spurious correlation in hydraulics and hydrology. *Journal of the Hydraulics Division of the American Society of Civil Engineers* 91: 35-42. **Biewener, A. A.** (2002). Walking with tyrannosaurs. *Nature* 415: 971-973. **Bland, R.** (1993). A closer look at AIDS dentist (letter). *The Wall Street Journal* June 28, 1993. **Blank, R. M.** (1991). The effects of double-blind versus single-blind reviewing: experimental evidence from The American Economic Review. *The American Economic Review* 81: 1041-1067. **Bodiselitsch, B.**, Koeberl, C., Master, S. and Reimold, W. U. (2005). Estimating duration and intensity of Neoproterozoic Snowball glaciations from Ir anomalies. *Science* 308: 239-242. **Bosch, R. A.** and Smith, J. A. (1998). Separating hyperplanes and the authorship of the disputed federalist papers. *American Mathematical Monthly* 105: 601–608. **Boutron, C. F.**, Görlach, U., Candelone, J., Bolshov, M. A. and Delmas, R. J. (1991). Decrease in anthropogenic lead, cadmium and zinc in Greenland snows since the late 1960s. *Nature* 353: 153-156. **Box, J. F. A.** (1978). *R. A. Fisher: the life of a scientist.* New York: Wiley. **Brand, L. R.** and Tang, T. (1991). Fossil vertebrate footprints in the Coconino Sandstone (Permian) of northern Arizona: evidence for underwater origin. *Geology* 19: 1201-1204. **Brecht, B. A.** (1993). *The Bertolt Brecht Journals 1934 to 1955.* New York: Routledge. **Brewer, G. D.** and Shubik, M. (1979). *The War Game: a critique of military problem solving.* Cambridge: Harvard University Press. **Bronowski, J.** (1965). *Science and Human Values.* New York: Harper & Row. **Browne, E. J.** (1995). *Voyaging. Charles Darwin, Volume 1.* New York: Knopf. **Brownell, P. H.** (1984). Prey detection by the sand scorpion. *Scientific American* 251(6): 86-97. **Burbidge, E. M.**, Burbidge, G. R., Fowler, W. A. and Hoyle, F. (1957). Synthesis of the elements in stars. *Reviews of Modern Physics* 29(4): 548-650. **Butler, J. C.** (1986). The role of spurious correlation in the development of a komatiite alteration model. (Proceedings of the seventeenth lunar and planetary science conference, part 1) *Journal of Geophysical Research* 91(B13): E275-E280. **Calvin, W. H.** (1983). *The Throwing Madonna: essays on the brain.* New York: McGraw-Hill. **Campbell, W. W.** (1908). The nature of an astronomer's work. *The North American Review* 187: 907-915. **Carroll, G. H.** (1933). Mist on the mirror. *The North American Review* 236(5): 448-455. **Caspi, A.**, Herbener, E. S. and Ozer, D. J. (1992). Shared experiences and the similarity of personalities: a longitudinal study of married couples. *Journal of Personality and Social Psychology* 62(2): 281-291. **Caws, P.** (1957). A reappraisal of the conceptual scheme of science. *Philosophy of Science* 24(3): 221-234. **Ceci, S. J.** (1991). How much does schooling influence general intelligence and its cognitive components? A reassessment of the evidence. *Developmental Psychology* 27(5): 703-722. **Chamberlin, T. C.** (1965). The method of multiple working hypotheses. *Science* 148: 754-759. **Chantrell, G.** (2002). *The Oxford Dictionary of Word Histories.* New York: Oxford University Press. **Chao, B. J.** (1995). Anthropogenic impact on global geodynamics due to reservoir water impoundment. *Geophysical Research Letters* 22: 3529-3532. **Chayes, F.** (1971). *Ratio Correlation: a manual for students of petrology and geochemistry.* Chicago: University of Chicago Press. **Chipp, H. B.** (1968). *Theories of Modern Art.* Berkeley: University of California Press. **Chitty, D.** (1967). The natural selection of self-regulatory behaviour in animal populations. *Proceedings of the Ecological Society of Australia.* 2: 51-73. **Choudhury, S.** (1995). Divorce in birds: a review of the hypotheses. *Animal Behaviour* 50: 413-429. **Chyba, C. F.** (1993). Explosions of small Spacewatch objects in the Earth's atmosphere. *Nature* 363: 701-703. **Cohen, B.** (2005). *The Triumph of Numbers: how counting shaped modern life.* New York: W. W. Norton & Company. **Colbeck, S. C.** (1995). Pressure melting and ice skating. *American Journal of Physics* 63(10): 888-890. **Connor, E. F.** and Simberloff, D. (1986). Competition, scientific method, and null models in ecology. *American Scientist* 74: 155-162. **Coppola, M.**

and Newport, E. L. (2005). Grammatical subjects in home sign: abstract linguistic structure in adult primary gesture systems without linguistic input. *Proceedings of the National Academy of Sciences of the United States of America* 102(52): 19249-19253. **Cousteau, J. Y.** and Schiefelbein, S. (2007). *The Human, the Orchid, and the Octopus: exploring and conserving our natural world.* New York: Bloomsbury. **Cowen, R.** (2006). Comet sampler. *Science News* 170: 387. **Cox, K. R.** (1968). *Planning Clinical Experiments.* Springfield, Ill.: Charles C. Thomas. **Croswell, K.** (2001). Wondering in the dark. *Sky & Telescope* 102(6): 44-50. **Davis, S.** and Payne, S. (1993). A barrow full of cattle skulls. *Antiquity* 67(254): 12-22. **Dawkins, R.** (1998). Science, delusion, and the appetite for wonder. *Skeptical Inquirer* 22(2): 28-&. **Dean, T. J.** and Cao, Q. V. (2003). Inherent correlations between stand biomass variables calculated from tree measurements. *Forest Science* 49(2): 279-284. **Dethier, V. G.** (1962). *To Know a Fly.* San Francisco: Holden-Day. **Dewdney, A. K.** (1997). *Yes, We Have No Neutrons: an eye-opening tour through the twists and turns of bad science.* New York: Wiley. **Drosnin, M.** (1997). *The Bible Code.* New York: Simon & Schuster. **Dyson, F. J.** (1979). *Disturbing the Universe.* New York: Basic Books. **Dyson, F. J.** (2004). One in a million. *The New York Review of Books* 51(5): 4-6. **Eberhardt, L. L.** (1970). Correlation, regression, and density dependence. *Ecology* 51(2): 306-310. **Eisenhart, C.** (1968). Expression of the uncertainties of final results: clear statements of the uncertainties of reported values are needed for their critical evaluation. *Science* 160: 1201-1204. **Enright, J. T.** (1999). Testing dowsing: the failure of the Munich experiments. *Skeptical Inquirer* 23(1): 39-46. **Esch, H. E.**, Zhang, S., Srinivasan, M. V. and Tautz, J. (2001). Honeybee dances communicate distances measured by optic flow. *Nature* 411: 581-583. **Faigman, D. L.** (2002). Science and the law: is science different for lawyers? *Science* 297: 339-340. **Few, A. A.** (1975). Thunder. *Scientific American* 233(5): 80-90. **Ford, K. W.** (2004). *The Quantum World.* Cambridge: Harvard University Press. **Franks, F.** (1981). *Polywater.* Cambridge: MIT Press. **Garcia, E.** and Carignan, R. (2005). Mercury concentrations in fish from forest harvesting and fire-impacted Canadian boreal lakes compared using stable isotopes of nitrogen. *Environmental Toxicology and Chemistry.* 24(3): 685-693. **Gardner, M.** (1986). *Knotted Doughnuts and Other Mathematical Entertainments.* New York: W. H. Freeman. **Gardner, M.** (2000). Little red riding hood. *Skeptical Inquirer* 24(5): 14-16. **Garfunkel, S.** (1987). *For All Practical Purposes: introduction to contemporary mathematics.* (Module 4: On size and shape). Arlington, MA: Consortium for Mathematics and its Applications. **Garsd, A.** (1984). Spurious correlation in ecological modeling. *Ecological Modelling* 23: 191-201. **Geedey, C. K.** and Dudycha, J. L. (2004). The stats guy. *Frontiers in Ecology and the Environment* 2(1): 49-50. **Gibbons, A.** (1997). Why life after menopause? *Science* 276: 535. **Goho, A.** (2005). Slick surfaces. *Science News* 167: 165-166. **Gomory, R. E.** (1983). Technology development. *Science* 220: 576-580. **Goodman, S.** and Greenland, S. (2007). Why most published research findings are false: problems in the analysis. *PLoS Medicine* 4(4): e168. **Gould, J. L.** (1985). How bees remember flower shapes. *Science* 227: 1492-1494. **Gould, S. J.** (1986). Entropic homogeneity isn't why no one hits 0.400 anymore. *Discover* 7(8): 60-66. **Gould, S. J.** (2001). A happy mystery to ponder: why so many homers? *The Wall Street Journal* October 10, 2001: A16. **Graveland, J.**, van der Wal, R., van Balen, J. H. and van Noordwijk, A. J. (1994). Poor reproduction in forest passerines from decline of snail abundance on acidified soils. *Nature* 368: 446-448. **Gu, J.-W.**, Bailey, A. P., Sartin, A., Makey, I. and Brady, A. L. (2005). Ethanol stimulates tumor progression and expression of vascular endothelial growth factor in chick embryos. *Cancer* 103: 422-431. **Guterman, L.** (2001). Verbatim. *The Chronicle of Higher Education* August 10, 2001: A12. **Haldane, J. B. S.** (1928). *Possible Worlds and Other Essays.* London: Chatto and Windus. **Hanson, N. R.** (1958). *Patterns of Discovery: an inquiry into the conceptual foundations of science.* New York: Cambridge University Press. **Hardarson, B. S.** and Fitton, J. G. (1991). Increased mantle melting beneath Snaefellsjökull volcano during Late Pleistocene deglaciation. *Nature* 353: 62-64. **Hardy, J.** and Selkoe, D. J. (2002). The amyloid hypothesis of Alzheimer's disease: progress and problems on the road to therapeutics. *Science* 297: 353-356. **Harris, J. E.**, Wente, E.F., Cox, C.F., El Nawaway, I., Kowalski, C. J., Storey, A. T., Russell, W.R., Ponitz, P.V, and Walker, G.F. (1978). Mummy of the "Elder Lady" in the tomb of Amenhotep II: Egyptian museum catalog number 61070. *Science* 200: 1149-1151. **Hasler, A. D.** (1966). *Underwater Guideposts: homing of salmon.* Madison: University of Wisconsin Press. **Hauck, S. A.**, Phillips, R. J. and Price, M. (1997). Implications for Venusian plains resurfacing from geomorphic mapping and impact crater densities. *Twenty-Eighth Annual Lunar and Planetary Science Conference.* Houston, TX, Lunar and Planetary Science. 28: 527-528. **Hawkes, K.** (2004). The grandmother effect. *Nature* 428: 128-129. **Hedge, F. H.** (1884). *Atheism in Philosophy, and Other Essays.* Boston: Roberts Brothers. **Heilbron, J. L.** (1999). *The Sun in the Church.* Cambridge: Harvard University Press. **Hemenway, D.**, Solnick, S. J., and Colditz, G. A. (1993). Smoking and suicide among nurses. *American Journal of Public Health* 83(2): 249-251. **Herbert, A.**, Gerry, N. P., McQueen, M. B., Heid, I. M., Pfeufer, A., Illig, T.,

Wichmann, H.-E., Meitinger, T., Hunter, D., Hu, F. B., Colditz, G., Hinney, A., Hebebrand, J., Koberwitz, K., Zhu, X., Cooper, R., Ardlie, K., Lyon, H., Hirschhorn, J. N., Laird, N. M., Lenburg, M. E., Lange, C. and Christman, M. F. (2006). A common genetic variant is associated with adult and childhood obesity. *Science* 312: 279 - 283. **Hibben, J. G.** (1908). The paradox of research. *The North American Review* 188: 425-431. **Hoffman, P. F.**, Kaufman, A. J., Halverson, G. P. and Schrag, D. P. (1998). A Neoproterozoic snowball Earth. *Science* 281: 1342-1346. **Hoffman, P. F.** and Schrag, D. P. (2000). Snowball Earth. *Scientific American* 282(1): 68-75. **Hofman, P. M.**, Van Riswick, J. G. and Van Opstal, A. J. (1998). Relearning sound localization with new ears. *Nature Neuroscience* 1(5): 417-421. **Holden, C.** (1996). Ebola: ancient history of "new" disease? *Science* 272: 1591. **Holton, G.** (1979). Constructing a theory: Einstein's model. *The American Scholar* 48: 309-340. **Hoyle, F.** (1950). *The Nature of the Universe.* New York: Harper. **Humphries, S.**, Bonser, R. H. C., Witton, M. P. and Martill, D. M. (2007). Did pterosaurs feed by skimming? Physical modelling and anatomical evaluation of an unusual feeding method. *PLoS Biology* 5(8): e240. **Hunter, G.** (2001). Little red riding hood, sex, symbols, and fairy tales? (letter). *Skeptical Inquirer* 25(1): 68. **Hutchinson, J. R.** and Garcia, M. (2002). *Tyrannosaurus* was not a fast runner. *Nature* 415: 1018-1021. **Huxley, J. S.** (1929). The size of living things. *The Atlantic Monthly* 144: 289-302. **Hyman, R.** (2005). Testing Natasha. *Skeptical Inquirer* 29(3): 27-31. **Ioannidis, J. P. A.** (2005). Why most published research findings are false. *PLoS Medicine* 2(8): e124. **Irion, R.** (2002). Rings of light could reveal black holes. *Science* 297: 2188. **Jablonski, D.** (1993). The tropics as a source of evolutionary novelty through geological time. *Nature* 364: 142-144. **Jackson, D. A.** and Somers, K. M. (1991). The spectre of 'spurious' correlations! *Oecologia.* 86: 147-151. **Jacob, F.** (1982). *The Possible and the Actual.* Seattle: University of Washington Press. **Jadad, A. R.** (1998). *Randomised Controlled Trials: a user's guide.* London: BMJ Books. **Jencks, C.** (2006). Letter. *The New York Review of Books* 53(6): 76. **Johannessen, O. M.**, Khvorostovsky, K., Miles, M. W. and Bobylev, L. P. (2005). Recent ice-sheet growth in the interior of Greenland. *Science* 310: 1013-1016. **Johnson, J. G.**, Cohen, P., Smailes, E. M., Kasen, S. and Brook, J. S. (2002a). Television viewing and aggressive behavior during adolescence and adulthood. *Science* 295: 2468-2471. **Johnson, J. G.**, Cohen, P., Smailes, E. M., Kasen, S. and Brook, J. S. (2002b). Letter. *Science.* 297: 50. **Kac, M.** (1983). Marginalia - statistical odds and ends. *American Scientist* 71: 186-187. **Kaufman, Y. J.** and Koren, I. (2006). Smoke and pollution aerosol effect on cloud cover. *Science* 313: 655-658. **Kellert, S. R.** and Wilson, E. O. (1993). *The Biophilia Hypothesis.* Washington, D.C.: Island Press. **Kemeny, J. G.** (1959). *Finite Mathematical Structures.* Englewood Cliffs, N.J.: Prentice-Hall. **Kenney, B. C.** (1982). Beware of spurious self-correlations. *Water Resources Research* 18(4): 1041-1048. **Kerr, R. A.** (1998). Global change: signs of past collapse beneath Antarctic ice. *Science* 281: 17. **Kerr, R. A.** (2000). Planetary science: Halley's origins mysterious no more? *Science* 290: 1071. **Kerr, R. A.** (2005). Cosmic dust supports a snowball earth. *Science* 308: 181. **Kirk, J.**, Ruiz, J., Chesley, J., Walshe, J. and England, G. (2002). A major Archean, gold- and crust-forming event in the Kaapvaal craton, South Africa. *Science* 297: 1856-1858. **Kite, G.** (1989). Some statistical observations. *Water Resources Bulletin* 25(3): 483-490. **Klar, A. J. S.** (2003). Human handedness and scalp hair-whorl direction develop from a common genetic mechanism. *Genetics* 165: 269-276. **Klarreich, E.** (2003). Bookish math: statistical tests are unraveling knotty literary mysteries. *Science News* 164: 392-394. **Kolata, G. B.** (1981). Clues to the cause of senile dementia. *Science* 211: 1032-1033. **Krutch, J. W.** (1948). *Henry David Thoreau.* New York: W. Sloane Associates. **Labarbera, M.** (1978). Particle capture by a Pacific brittle star: experimental test of the aerosol suspension feeding model. *Science* 201: 1147-1149. **Lacina, A.** (1999). Atom – from hypothesis to certainty. *Physics Education* 34(6): 397-402. **Lahdenperä, M.**, Lummaa, V., Helle, S., Tremblay, M. and Russell, A. F. (2004). Fitness benefits of prolonged post-reproductive lifespan in women. *Nature* 428: 178-181. **Leather, J.**, Allen, P. A., Brasier, M. D. and Cozzi, A. (2002). Neoproterozoic snowball Earth under scrutiny: evidence from the Fiq glaciation of Oman. *Geology* 30(10): 891-894. **Lee, H. D. P.** (1952). *Aristotle Meteorologica.* London: William Heinemann Ltd. **Letters-to-the-editor** (1999). Evaluating the Munich dowsing experiments. *Skeptical Inquirer* 23(3): 64-65. **Lewis, D. W.** (1992). What was wrong with Tiny Tim? *American Journal of Diseases of Children* 146(12): 1403-1407. **Lincoln, H.** (1991). The holy place (letter). *The Times Literary Supplement,* April 12, 1991. **Lind, J.** (1753). *A treatise of the scurvy. In three parts.* London: A. Millar. **Linde, A. T.** and Sacks, I. S. (1998). Triggering of volcanic eruptions. *Nature* 395: 888-890. **Lipton, P.** (2005). Testing hypotheses: prediction and prejudice. *Science* 307: 219-221. **Livingston, R.**, Adam, B. S. and Bracha, H. S. (1993). Season of birth and neurodevelopmental disorders: summer birth is associated with dyslexia. *Journal of the American Academy of Child and Adolescent Psychiatry* 32(3): 612-616. **Loftus, L.S.** and Arnold, W.N. (1991). Vincent van Gogh's illness: acute intermittent porphyria. *British Medical Journal* 303: 1589-1591. **Luria, S. E.** (1984). *A Slot Machine, a Broken Test Tube:*

an autobiography. New York: Harper & Row. **Macdonald, A. B.** (1930). Article republished as "A Kansas City Star reporter turns detective and solves a murder in Amarillo" *in:* L. L. Snyder and R. B. Morris, eds. *A Treasury of Great Reporting.* (2nd edition, 1962). New York: Simon and Schuster: 471-475. **Maddox, B.** (2000). Murder, she wrote. *New York Times Book Review* April 23, 2000: 14. **Madsen, K. M.**, Lauritsen, M. B., Pedersen, C. B., Thorsen, P., Plesner, A., Andersen, P. H. and Mortensen, P. B. (2003). Thimerosal and the occurrence of autism: negative ecological evidence from Danish population-based data. *Pediatrics* 112(3): 604-606. **Madsen, T.**, Shine, R., Loman, J. and Håkansson, T. (2002). Why do female adders copulate so frequently? *Nature* 355: 440-441. **Mahlman, J. D.** (1997). Uncertainties in projections of human-caused climate warming. *Science* 278: 1416 - 1417. **Manyande, A.**, Chayen, S., Priyakumar, P., Smith, C. C., Hayes, M., Higgins, D., Kee, S., Phillips, S. and Salmon, P. (1992). Anxiety and endocrine responses to surgery: paradoxical effects of preoperative relaxation training. *Psychosomatic Medicine* 54(3): 275-287. **Marden, J. H.** and Kramer, M. G. (1994). Surface-skimming stoneflies: a possible intermediate stage in insect flight evolution. *Science* 266: 427-430. **Margenau, H.** (1961). *Open Vistas: philosophical perspectives of modern science.* New Haven: Yale University Press. **Martin, B.** (1998). Coincidences: remarkable or random? *Skeptical Inquirer* 22(5): 23-28. **Marzuola, C.** (2002). Snowball melting? *Science News* 162: 246. **Mathews, G. J.** and Schramm, D. N. (1993). Protogalactic mergers and cosmochronology. *The Astrophysical Journal* 404(1): 468-475. **McCuen, R. H.** (1974). Spurious correlation in estimating recreation demand functions. *Journal of Leisure Research* 6: 232-240. **McGue, M.** and Lykken, D. T. (1992). Genetic influence on risk of divorce. *Psychological Science* 3(6): 368-373. **Medawar, P. B.** (1969). *Induction and Intuition in Scientific Thought.* Philadelphia: American Philosophical Society. **Meylan, A. B.**, Bowen, B. W. and Avise, J. C. (1990). A genetic test of the natal homing versus social facilitation models for green turtle migration. *Science* 248: 724-727. **Mills, L. S.** (2007). *Conservation of Wildlife Populations: demography, genetics, and management.* Malden, MA: Blackwell Publishing. **Molesworth, W.** (1845). *The English Works of Thomas Hobbes* (Vol VII). London: John Bohn. **Monastersky, R.** (2004). The cold, cold war: geologists heatedly debate whether the earth ever froze over completely. *The Chronicle of Higher Education* January 9, 2004: A12-A14. **Morse, D. H.** (1985). Milkweeds and their visitors. *Scientific American* 253(1): 112-119. **Mosey, N. J.**, Müser, M. H. and Woo, T. K. (2005). Molecular mechanisms for the functionality of lubricant additives. *Science* 307: 1612-1615. **Mosteller, F.** and Wallace, D. L. (1964). *Inference and Disputed Authorship: The Federalist.* Reading, Mass.: Addison-Wesley. **Muller, R. A.** (1985). An adventure in science. *The New York Times Magazine* March 24, 1985: 34-&. **Müller, W.**, Fricke, H., Halliday, A. N., McCulloch, M. T. and Wartho, J. (2003). Origin and migration of the alpine iceman. *Science* 302: 862-866. **Nadis, S.** (2002). Using lasers to detect E.T. *Astronomy* 30(9): 45-48. **Naqvi, N. H.**, Rudrauf, D., Damasio, H. and Bechara, A. (2007). Damage to the insula disrupts addiction to cigarette smoking. *Science* 315: 531-534. **Nazé, Y.** (2006). Live fast, die young. *Astronomy* 34(4): 40-45. **Newman, J. R.** (1956). *The World of Mathematics.* (Volume 1). New York: Simon and Schuster. **Newman, M. F.**, Kirchner, J. L., Phillips-Bute, B., Gaver, V., Grocott, H., Jones, R. H., Mark, D. B., Reves, J. G. and Blumenthal, J. A. (2001). Longitudinal assessment of neurocognitive function after coronary-artery bypass surgery. *The New England Journal of Medicine* 344: 395-402. **Nof, D.** and Paldor, N. (1992). Are there oceanographic explanations for the Israelites' crossing of the Red Sea? *Bulletin of the American Meteorological Society* 73(3): 305-314. **O'Keefe, J. A.** and Glass, B. P. (1985). Lunar sample-14425: characterization and resemblance to high-magnesium microtektites. *Science* 227: 515-516. **Olson, D. W.**, Doescher, R. L., Burke, A. K., Delgado, M. E., Douglas, M. A., Fields, K. L., Fischer, R. B., Gardiner, P. D., Huntley, T. W., McCarthy, K. E. and Messenger, A. G. (1994). Dating Ansel Adam's *Moon and Half Dome. Sky & Telescope* 88(6): 82-86. **Owen, R.** (1982). Reader bias. *The Journal of the American Medical Association* 247(18): 2533-2534. **Parker, P.** (1997). Right lung, wrong count. *Science News* 152: 56. **Partridge, B. L.** (1982). The structure and function of fish schools. *Scientific American* 114-123. **Payne, S.** (1974). Terrestrial model: animal processes (version III). *US/IBP Desert Biome Research Memorandum 74-52.* Utah State University: 74-152. **Penrose, R.** (1997). The mathematics of the electron's spin. *European Journal of Physics* 18(3): 164-168. **Perkins, S.** (2001). Big dam in China may warm Japan. *Science News* 159: 245. **Perry, G.**, Raina, A. K., Cohen, M. L. and Smith, M. A. (2004). When hypotheses dominate. *The Scientist* 18(23): 6. **Petry, D.** (1992). Landslide theories (letter). *Science News* 142: 267. **Phillips, D. P.** (1978). Airplane accident fatalities increase just after newspaper stories about murder and suicide. *Science* 201: 748-750. **Phillips, D. P.** and King, E. W. (1988). Death takes a holiday: mortality surrounding major social occasions. *Lancet* 2(8613): 728-732. **Phillips, D. P.** and Smith, D. G. (1990). Postponement of death until symbolically meaningful occasions. *The Journal of the American Medical Association* 263(14): 1947-1951. **Phillips, D. P.**, Van Voorhees, C. A. and Ruth, T. E. (1992). The birthday:

lifeline or deadline? *Psychosomatic Medicine* 54(5): 532-542. **Phillips, P. C. B.** (1986). Understanding spurious regressions in economics. *Journal of Econometrics* 33: 311-340. **Platt, J. R.** (1964). Strong inference. *Science* 146: 347-353. **Posner, G. P.** (2000). The face behind the "face" on Mars: a skeptical look at Richard C. Hoagland. *Skeptical Inquirer* 24(6): 20-26. **Power, M. E.** (1990). Effects of fish in river food webs. *Science* 250: 811-814. **Profet, M.** (1993). Menstruation as defense against pathogens transported by sperm. *The Quarterly Review of Biology* 68(3): 335-386. **Prusiner, S. B.** (1998). Prions. *Proceedings of the National Academy of Sciences* 95: 13363-13383. **Raloff, J.** (2008). Judging science. *Science News* 173: 42-44. **Ray, T. P.** (1989). The winter solstice phenomenon at Newgrange, Ireland: accident or design? *Nature* 337: 343-345. **Richards, D. N.** (1992). Landslide theories (letter). *Science News* 142: 267. **Ridley, M.** (2000). How far from the tree? *New York Times Book Review* August 20, 2000: 14. **Rodbard, D.** (1982). A by-line for statistical consultants? *The American Statistician* 36(3): 217-218. **Romesburg, H. C.** (1979). Simulating scientific inquiry with the card game Eleusis. *Science Education* 63(5): 599-608. **Romesburg, H. C.** (1981). Wildlife science: gaining reliable knowledge. *The Journal of Wildlife Management* 45(2): 293-313. **Romesburg, H. C.** (1985). Exploring, confirming, and randomization tests. *Computers & Geosciences* 11(1): 19-37. **Romesburg, H. C.** (1989). ZORRO: a randomization test for spatial pattern. *Computers & Geosciences* 15(6): 1011-1017. **Romesburg, H. C.** (1991). On improving the natural resources and environmental sciences. *The Journal of Wildlife Management* 55: 1177-1180. **Romesburg, H. C.** (1993). On improving the natural resources and environmental sciences: a reply. *The Journal of Wildlife Management* 57(1): 184-189. **Romesburg, H. C.** (2001). *The Life of the Creative Spirit*. Philadelphia: Xlibris Corporation. **Romesburg, H. C.** (2004). *Cluster Analysis for Researchers*. Morrisville, NC: Lulu.com. **Romesburg, H. C.** (2005). *How About it, Writer?* Morrisville, NC: Lulu.com. **Romm, J.** (1994). A new forerunner for continental drift. *Nature* 367: 407-408. **Rosenbaum, L.** (1996). Art: Michelangelo (?) in America. *The Wall Street Journal* February 29, 1996: A16. **Rozin, P.**, Kabnick, K., Pete, E., Fischler, C. and Shields, C. (2003). The ecology of eating: smaller portion sizes in France than in the United States helps explain the French paradox. *Psychological Science* 14(5): 450-454. **Ruscio, J.** (2005). Exploring controversies in the art and science of polygraph testing. *Skeptical Inquirer* 29(1): 34-39. **Russell, B.** (1974). *The Art of Philosophizing and Other Essays*. Totowa, N.J: Littlefield, Adams. **Saila, S. B.** and Shappy, R. A. (1962). Migration by computer. *Discovery* 23(6): 23-26. **Sanchez-Levega, A.**, Colas, F., Lecacheux, J., Laques, P., Miyazaki, I. and Parker, D. (1991). The great white spot and disturbances in Saturn's equatorial atmosphere during 1990. *Nature* 353: 397-401. **Sayers, D. L.** (1941). *The Mind of the Maker*. New York: Harcourt, Brace and Co. **Schaefer, J. M.**, Denton, G. H., Barrell, D. J. A., Ivy-Ochs, S., Kubik, P. W., Andersen, B. G., Phillips, F. M., Lowell, T. V. and Schlüchter, C. (2006). Near-synchronous interhemispheric termination of the last glacial maximum in mid-latitudes. *Science* 312: 1510-1513. **Schilling, G.** (2003). Liquid-mirror telescope set to give stargazing a new spin. *Science* 299: 1650. **Schilling, G.** (2006). Do gamma ray bursts always line up with galaxies? *Science* 313: 749. **Scholz, A. T.**, Horrall, R. M., Cooper, J. C. and Hasler, A. D. (1976). Imprinting to chemical cues: the basis for home stream selection in salmon. *Science* 192: 1247-1249. **Schwenk, K.** (1994). Why snakes have forked tongues. *Science* 263: 1573-1577. **Scott, E. L.** (1979). Correlation and suggestions of causality: spurious correlation. Pages 237-251 *in* L. Orloei, C. R. Rao, and W. M. Stiteler, eds. *Multivariate Methods in Ecological Work*. Fairland, Maryland: International Co-operative Publishing House. **Scott, S.** and Duncan, C. J. (2001). *Biology of Plagues: evidence from historical populations*. Cambridge: Cambridge University Press. **Seife, C.** (2004). Physics enters the twilight zone. *Science* 305: 464-466. **Selnes, O. A.** and McKhann, G. M. (2001). Coronary-artery bypass surgery and the brain. *The New England Journal of Medicine* 344: 451-452. **Shearer, R. A.** (1999). Statement analysis: SCAN or scam. *Skeptical Inquirer* 23(3): 40-43. **Shettleworth, S. J.** (1983). Memory in food-hoarding birds. *Scientific American* 248(3): 102-110. **Siegel, J. M.** (1990). Stressful life events and use of physician services among the elderly: the moderating role of pet ownership. *Journal of Personality and Social Psychology* 58(6): 1081-1086. **Simkin, B.** (1992). Mozart's scatological disorder. *British Medical Journal* 305(6868): 1563–1567. **Simon, H. A.** (1991). *Models of My Life*. New York: Basic Books. **Skolnick, A. A.** (2005). Natasha Demkina: the girl with normal eyes. *Skeptical Inquirer* 29(3): 34-37. **Small, G. W.**, Propper, M. W., Randolph, E. T. and Eth, S. (1991). Mass hysteria among student performers: social relationship as a symptom predictor. *American Journal of Psychiatry* 148(9): 1200-1205. **Sockloff, A. L.** (1976). Spurious product correlation. *Educational and Psychological Measurement* 36: 33-44. **Srygley, R. B.** (2001). Sexual differences in tailwind drift compensation in *Phoebis sennae* butterflies (Lepidoptera: Pieridae) migrating over seas. *Behavioral Ecology* 12(5): 607-611. **Stern, K.** and McClintock, M.K. (1998). Regulation of ovulation by human pheromones. *Nature* 392:177. **Stewart, I.** (1998). Counting the pyramid builders. *Scientific American* 279(3): 98-100. **Stokstad, E.** (2002). *T. rex* was no runner, muscle study shows. *Science* 295: 1620-1621. **Stone, E. M.**, Nichols, B. E., Streb, L. M., Kimura, A. E. and Sheffield, V. C. (1992). Genetic linkage of vitelliform macular degeneration (Best's disease) to chromosome 11q13. *Nature Genetics* 1(4): 246-250. **Stone, R.** (2008). Preparing for doomsday. *Science* 319: 1326-1329. **Strassmann, B. I.** (1996). The evolution of endometrial cycles and menstruation. *The Quarterly Review of Biology* 71(2): 181-220. **Swartz, C. E.** (2003). *Back-of-the-Envelope Physics.* Baltimore: Johns Hopkins University Press. **Swenson, G. W. Jr.** (2000). Searching for extraterrestrials. Intragalactically speaking. *Scientific American* 283(1): 44-47. **Szibor, R.**, Schubert, C., Schöning, R., Krause, D. and Wendt, U. (1998). Pollen analysis reveals murder season. *Nature* 395: 449-450. **Szilard, L.** (1961). *The Voice of the Dolphins, and Other Stories.* New York: Simon and Schuster. **Templeton, J. M.** (1997). How TWA 800 may have exploded (letter). *The Wall Street Journal* December 15, 1997. **Tennekes, H.** (1996). *The Simple Science of Flight: from insects to jumbo jets.* Cambridge: MIT Press. **Thomas, D. E.** (1997). Hidden messages and the Bible code. *Skeptical Inquirer* 21(6): 30-36. **Thomson, K. S.** (1997). Natural selection and evolution's smoking gun. *American Scientist* 85: 516-518. **Trefil, J.** (1998). How stars shine. *Astronomy* 26(1): 56-60. **Trefil, J.** (2000). Putting stars in their place. *Astronomy* 28(11): 62-67. **Trefil, J.** (2005). Relativity's infinite beauty. *Astronomy* 33(2): 46-53. **Turner, M. D.** and Rabinowitz, D. (1983). Factors affecting frequency distributions of plant mass: the absence of dominance and suppression in competing monocultures of *Festuca Paradoxa. Ecology* 64(3): 469-475. **Turner, M. S.** (2007). Quarks and the cosmos. *Science* 315: 59-61. **Twain, M.** (1996). *1601, and Is Shakespeare Dead?* (S. F. Fishkin ed.) New York: Oxford University Press. **Tyson, N. D.** (1989). *Merlin's Tour of the Universe.* New York: Columbia University Press. **Tyson, N. D.** (2003). *My Favorite Universe* (course guidebook). Chantilly, Va.: The Teaching Company. **Vander Wall, S. B.** (1982). An experimental analysis of cache recovery in Clark's nutcracker. *Animal Behaviour* 30: 84-94. **Vergano, D.** (1996). Smallest frog leaps into the limelight: eleuth frog discovered in Cuban rain forest. *Science News* 150: 357. **Vogel, S.** (1988). *Life's Devices: the physical world of animals and plants.* Princeton: Princeton University Press. **Vogel, S.** (1998). *Cats' Paws and Catapults: mechanical worlds of nature and people.* New York: W. W. Norton & Company. **Vollrath, F.** (1979). Vibrations: their signal function for a spider kleptoparisite. *Science* 205: 1149-1151. **Walker, H. M.** (1939). Testing a statistical hypothesis. *Harvard Educational Review* 9(2): 229-240. **Wang, F.**, Garza, L. A., Kang, S., Varani, J., Orringer, J. S., Fisher, B. J. and Voorhees, J. J. (2007). In vivo stimulation of de novo collagen production caused by cross-linked hyaluronic acid dermal filler injections in photodamaged human skin. *Archives of Dermatology* 143: 155-163. **Wannamethee, S. G.**, Shaper, A. G., Lennon, L. and Whincup, P. H. (2006). Height loss in older men: associations with total mortality and incidence of cardiovascular disease. *Archives of Internal Medicine* 166: 2546-2552. **Weaver, W.** (1982). *Lady Luck: the theory of probability.* New York: Dover. **Weinberg, S.** (1990). Steven Weinberg interview. *in:* A. P. Lightman and R. Brawer, eds. *Origins: the lives and worlds of modern cosmologists.* Cambridge: Harvard University Press: 451-466. **Weinberg, S.** (2003). *The Discovery of Subatomic Particles: revised edition.* Cambridge: Cambridge University Press. **Weisberg, S.** (1980). *Applied Linear Regression.* New York: Wiley. **Weiss, J.** (1990). Unconscious mental functioning. *Scientific American* 262(3): 103-109. **Weldon, P. J.**, Demeter, B. J. and Rosscoe, R. (1993). A survey of shed skin-eating (dermatophagy) in amphibians and reptiles. *Journal of Herpetology* 27(2): 219-228. **Went, F. W.** (1968). The size of man. *American Scientist* 56: 400-413. **Whitehead, H.** (1985). Why whales leap. *Scientific American* 252(3): 84-93. **Wiley, S.** (2008). No to negative data. *The Scientist* 22(4): 39. **Wilson, P. W.** (1930). Curse or coincidence? *The North American Review* 230: 85-92. **Wittlinger, M.**, Wehner, R. and Wolf, H. (2006). The ant odometer: stepping on stilts and stumps. *Science* 312: 1965-1967. **Woodard, G. C.** (1992). Landslide theories (letter). *Science News* 142: 267. **Youden, W. J.** (1972). Enduring values. *Technometrics* 14(1): 1-11. **Zee, A.** (1989). *An Old Man's Toy.* New York: Macmillan. **Zimmer, C.** (2004). Human evolution: faster than a hyena? Running may make humans special. *Science* 306: 1283.

Index

Subjects of research examples, in page order: